With Scott Before the Mast
by
Francis Davies

British Antarctic Expedition
1910 - 1913

These are the Journals of Francis Davies
Leading Shipwright RN when on board
Captain Scott's Terra Nova

Published by
Reardon Publishing
PO Box 919, Cheltenham, UK, GL50 9AN
www.reardon.biz

ISBN 9781901037555 (Hardback Edition)
ISBN 9781901037692 (Special Limited Edition)

Written by Francis Davies
Copyright 2020

Edited by Joy Watts
With thanks to everyone who has helped in this enjoyable endeavour to bring Chippy's journal to life.

Book Design by Nicholas Reardon

***

Endpaper Front:
Seamen crew of the Terra Nova 1910
(Left to right)

Back rows : Horton, Brissenden, Parsons, Heald, Neale, Balson, McCarthy
Centre rows: A.McDonald, W.McDonald, Burton, McGillon, McKenzie, Omenchelco, Clissold, Mather, Davies
Front rows: Skelton, McLoed, Bailey, Forde, Leese

Endpaper Back:
Meares and Oates at the blubber stove in the stables.

Foreword by David Wilson

The geographic and scientific accomplishments of Captain Scott's two Antarctic expeditions changed the face of the Twentieth Century in ways that are still not widely appreciated over a hundred years later. The fact of accomplishment has tended to be lost in speculative argument as to how Scott should have done this instead of that, supposedly to achieve the extra few yards per day to save the lives of the South Pole Party in 1912. Also lost to a generation overwhelmed with information, however, is the sublime sense of adventure into the unknown, which Scott's expeditions represented to his generation. We have forgotten what it is to take the awesome life-gambling risk of sailing beyond the edge of the map into nothingness and rendering it known. We send robot explorers instead. As a result, after two millennia of maritime and exploration history, we have become detached from the sea which surrounds our island and the tradition of exploration which it represents.

With Scott: Before the Mast is a unique account that serves as an antidote to this disconectedness. It is no fictional 'Hornblower', although it may seem so at times. This is a true story. It presents one man's account of his part in a great act of derring-do, the assault on the South Pole in 1912. Most records of Captain Scott's British Antarctic Expedition aboard *Terra Nova* (1910-1913) are the accounts of officers. *With Scott: Before the Mast* is the story of Francis Davies, Shipwright RN and Carpenter. The title says it all but may be lost on landlubbers. *Before the Mast* means 'to serve as an ordinary seaman in a sailing ship'. This makes it a rare and hugely important account, presenting a viewpoint from the lower ranks. Such insight is rarely available and the long overdue publication of this account is greatly to be welcomed.

When I first read this manuscript some years ago, I was hugely excited by the refreshing perspective that it gave to a well-aired story. Although an autobiographical period piece, written with an eye to publication many years after the events that it recalls, it is still of great interest. It tells the often forgotten story of the vast majority of Scott's men, the sailors of *Terra Nova*; the supporting cast, if you like, to the Shore Parties of officers and scientists. Through a kaleidoscope of memories, this book gets to the heart of the huge logistic effort that was the British Antarctic Expedition. Through the eyes of the sailors who were its backbone, the heroism of derring-do seems to become all the more human, humdrum and every-day. To me, this makes the accomplishments all the more real and remarkable.

I commend this book to collectors, enthusiasts, historians and to those looking for a casual read on Captain Scott, in equal measure.

David M. Wilson
Polar Historian and Gt. Nephew, Edward Wilson of the Antarctic

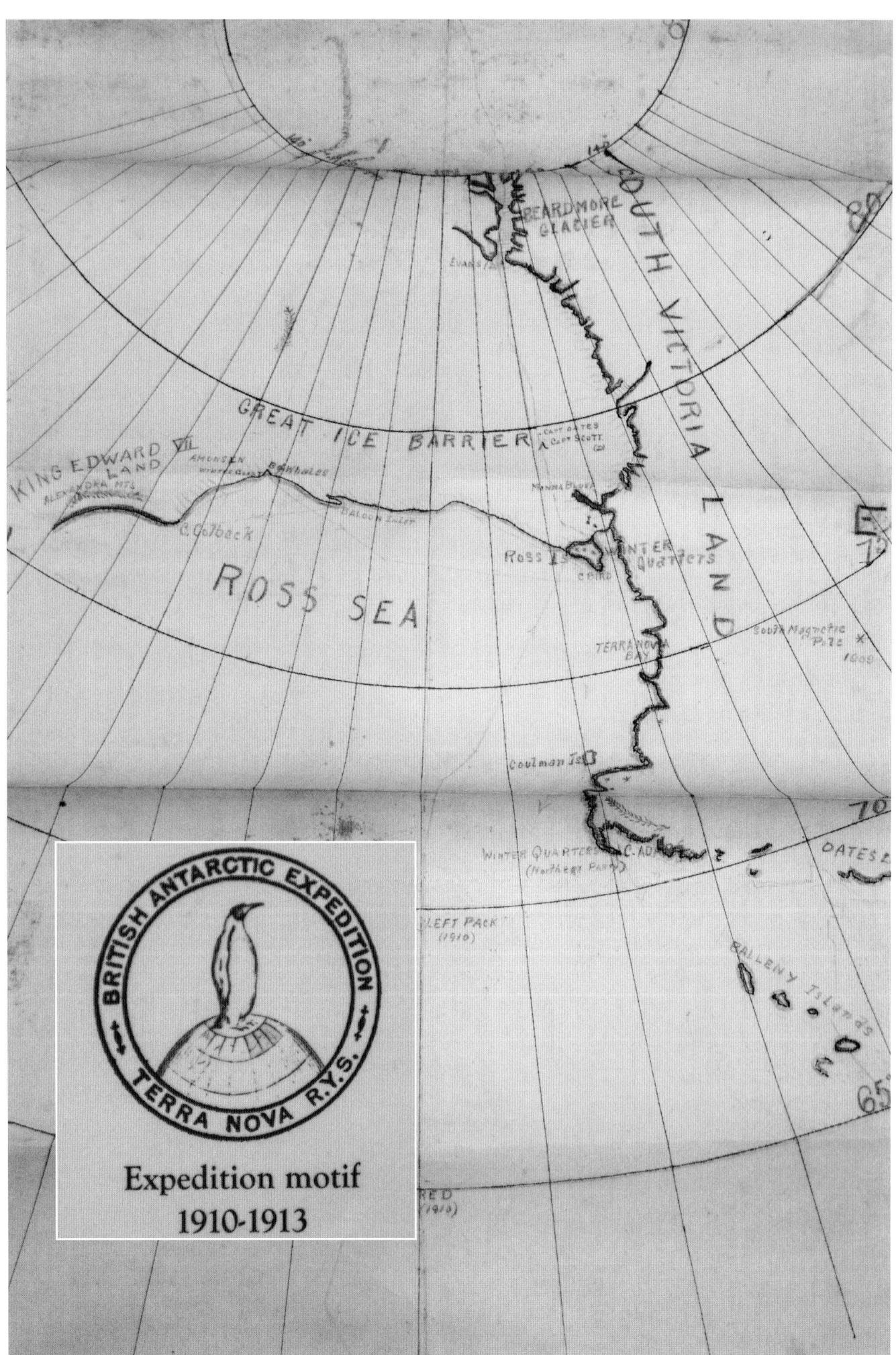

Expedition motif 1910-1913

Preface by Joy Watts great niece of Francis Davies

When I was a child, my mother often spoke of Great Uncle Frank's Antarctic adventures. His daughter, my cousin Beatrice, always known to the family as Maidie was custodian of her father's trunk holding artefacts and souvenirs from his polar travels. I refer to Francis Davies, Leading Shipwright RN, and Carpenter on the *Terra Nova* British Antarctic Expedition 1910 - 1913.

Occasionally, Maidie would open the trunk and we would peer at the objects before us. A diverse collection, among it penguins' eggs, lava from Mt. Erebus, charts and photographs, diaorami, tools and items used on *Terra Nova* and most intriguing to me at the time, a jar containing a seal embryo preserved in spirit. Also in the trunk and most treasured by Maidie was her father's account of his expedition experiences which he named 'With Scott: Before The Mast' written under the pseudonym Rudolph.

Francis Davies was born in Plymouth in 1885, where he was brought up in the Lower Crab Tree and Laira Green area of the city, attending Laira Green School. On leaving school and prior to his entry into the Royal Navy, he attained a shipwright's apprenticeship at the Royal Naval Dockyard, Devonport. It was as Shipwright on HMS *Vanguard* and when working in Devonport, he first heard mention of Captain Scott's plans for an expedition to the Antarctic and that shipwrights were required. Such was his longing for adventure into the unknown, he immediately knew he wanted to be a part of it. He applied, was accepted and duly appointed Leading Shipwright. He joined the British Antarctic Expedition on 30 May 1910 and signed on at Poplar, London.

Mustard Pot used on Terra Nova.

Often called Chippy by his shipmates his many skills were always in demand and constantly put to the test both on the voyage and ashore. Beginning with his work in the refit of *Terra Nova* from a blubber laden whaler, in a very poor state, to an expedition ship. He was meticulous in his most principal task, this being his planning and building of the huts, both the living quarters at Cape Evans for the Southern Party, about which Captain Scott wrote 'we are simply overwhelmed by it's comfort' and eventually at Cape Adare for the now Northern Party. The latter having to be rapidly constructed due to the urgency for *Terra Nova* to leave for New Zealand. Notably, he played a significant part in the emergency work in the engine room during the storm in the Southern Ocean at the very start of the expedition, when all was threatened and disaster narrowly averted. Along the way he was witness to the extraordinary meeting of *Terra Nova* with Amundsen's ship *Fram* in the Bay of Whales. To his final Antarctic task, the construction of the Memorial Cross to commemorate Captain Scott and the South Pole Party who perished on their return journey from the Pole. He chose to use the extremely hard Australian jarra wood. The Cross stands to this day on the top of Observation Hill overlooking McMurdo Sound.

Scott's hut at Cape Evans also still stands, both structures are designated Antarctic Historic Monuments. Davies' name is commemorated by Davies Bay, situated between Drake Head and Cape Kinsey, which was discovered in February 1911.

After the expedition Francis Davies served in the First World War and in 1920 took early voluntary retirement from the Royal Navy. He also served on Royal Research Ships *Discovery II* and *William Scorseby* which were engaged in scientific work in the Southern Ocean regions. He later volunteered and served in the Second World War. All his career he sailed on long voyages often lasting years but always returned to the Plymouth area, Drake's country as he so fondly called it. He married Ethel Stephens and then lived at Nicholls Farm, Plympton with their children, Beatrice and Peter. Francis Davies died in Plymouth in 1952, his ashes being scattered on the sea in sight of the National Memorial to Captain Scott and the Polar Party at Mount Wise, Devonport.

My sincere thanks to all who have supported me with this publication. To... David Wilson for his initial encouragement and forward. Robert Headland of the Scott Polar Research Institute for kindly agreeing to proof read. Paul Davies, President of the Devon and Cornwall Polar Society, for his help and advice, and to fellow members of the society including, Michael Tarver and Julie Ellis for their enthusiasm. It was sometime ago when David first introduced me to my patient publisher, Nicholas Reardon, together, 'With Scott: Before the Mast' has at long last been printed.

I dedicate this narrative of adventure to the memory of my cousin Maidie, who first introduced me to the wonders of the heroic age of polar exploration, and those who dared.

I hope you enjoy Chippy's tale.
Joy Watts.

A sketch by Francis Davies of his home Nicholls Farm, Plympton.

Prologue

Exploration !

The magic word that dominated my early life and filled my mind with visions of the first primitive, rough-hewn boats, then galleys, followed by boats with coarse sails, graceful sailing-ships with wind-filled canvas, steamships throbbing with speed, and grim, grey warships relentless in their messiness of unsinkable armoury.

It was not, however, the ships that captured my youthful imagination. It was the distant lands they were bound for.

I wanted to see immeasurable stretches of ocean swirling in unsounded deeps. I yearned to peer into limitless gulfs of space, to gaze across thousands of miles of silent, frozen vastness constituting the Unknown – desolate immensity shrouded in eternal night.

I wanted to contact the Antarctic.

Francis Davies

Francis Davies Leading Shipwright RN

The ship was rolling drunkenly, when one of the doctors and I fetched a small downton pump from for'd, to use in the engine-room. We had struggled along the deck as far as the break of the poop when a big sea came over the rails and we floated away, pump and all, rolling from side to side of the ship, until we were grabbed by some of the hands manning the hand pump. When we eventually got the pump to the stokehold it was of little use.

Captain Scott and Lieut Evans discussed the situation and came to the conclusion the only thing to be done was to cut a hole through the steel bulkhead between the boiler-room and the main hold, in order to get at the hand pump suctions, and in the meantime to bail out with buckets. Fortunately I had just the tool for the job – a drivis for cutting steel plate. Bill Lashly and myself commenced to cut the hole immediately, lying on the...

the Journals of Francis Davies

Lieut Evans went in first and I followed. There was so much water that it was difficult to reach the bottoms of the suctions; we practically had to dive under water. When we had cleared a certain amount of the muck away, we put rat traps under the suction pipes; these helped, greatly, to prevent them choking so often.

When the suctions were cleared I went on deck and stripped the pump to clean all the valves and once again it was under way, going full bore. By the time the pump was O.K. it was about midnight and I thought I would try and get a couple of hours 'shut eye'. As I went for'd to my mess, I looked into the foc'sle to see how Captain Oates and Doctor Atkinson were faring with the ponies.

They were having a very bad time indeed, two of the ponies were already down.

Captain Oates asked if I could give them a hand to put canvas slings under them to try and keep them on their legs. This was no easy matter as there was little or no room to move and the ponies on either side nipped us with their teeth whenever they got the chance. However, we got them

Francis Davies with his wife Ethel

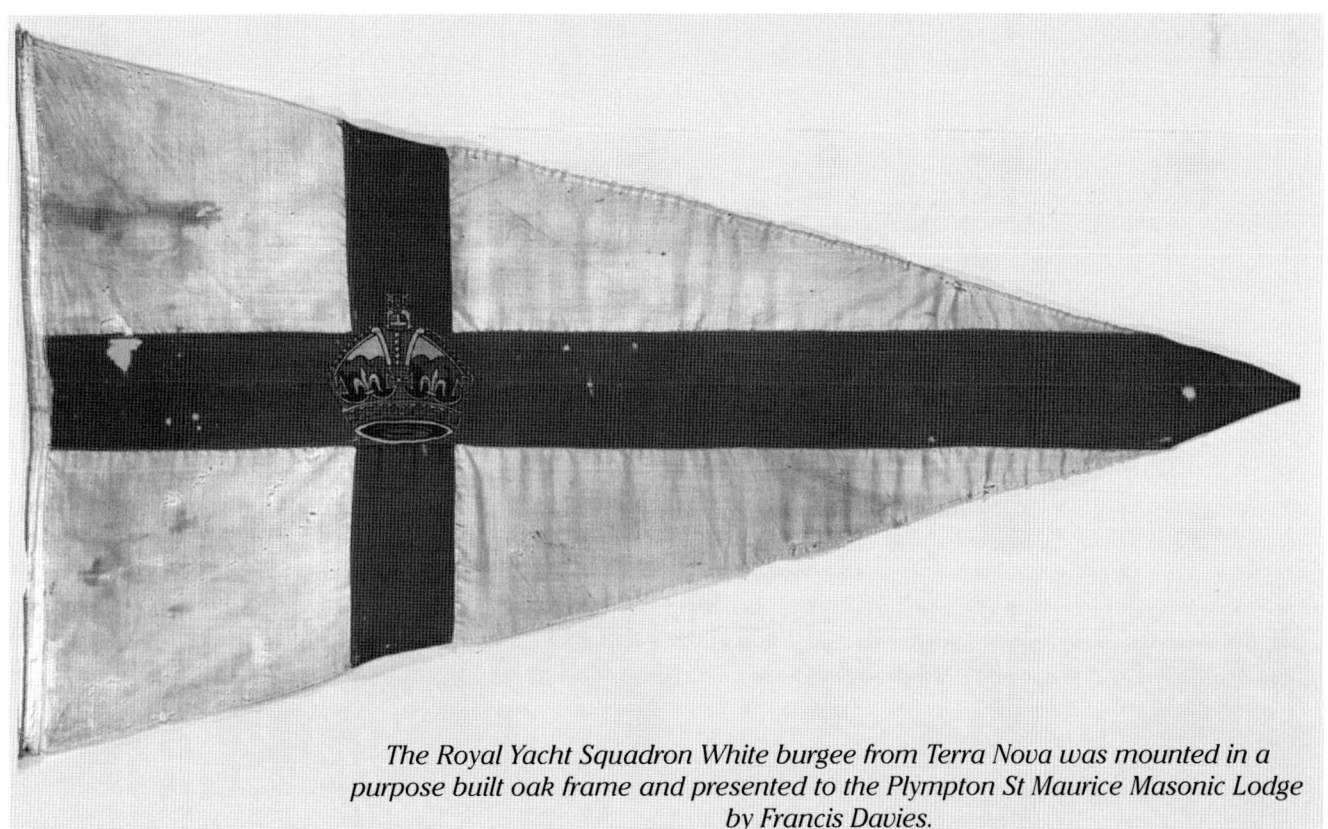

The Royal Yacht Squadron White burgee from Terra Nova was mounted in a purpose built oak frame and presented to the Plympton St Maurice Masonic Lodge by Francis Davies.
It was auctioned at Christie's in 1999.
It is now in the National Maritime Museum at Greenwich.

Chapter I. Signing on

For over nine years from first to last, I served in expedition ships engaged in exploration and scientific research in the Antarctic.

My first ship was *Terra Nova* of the British Antarctic Expedition 1910 – Scott's Last Expedition, and it is of my experiences, generally, in this expedition that I am writing. Looking back, I now see it was the end of an era in Antarctica or more correctly perhaps, of Polar exploration when the work was carried on in wooden sailing ships of great strength, especially constructed to withstand ice pressure, and with auxiliary steam power for working through the heavy ice-floes. The vessels were built for whaling and sealing in the Polar Regions, and were the pride of the Dundee shipbuilders during the latter half of the last century.

My association with exploration started in the spring of 1910 when I was serving as a shipwright in HMS *Vanguard*, Super-Dreadnought, recently commissioned for the first time. One day, whilst my mate and I were in the dockyard at Devonport scrounging material for a particular job we had in hand, he met one of his old shipmates who during a 'quack' about old times, mentioned he had heard that three shipwrights were required as volunteers for an expedition to the Antarctic, to be led by Captain Scott. I immediately cocked up my ears and from a few apparently disinterested remarks, I gathered that the carpenter who had served with Captain Scott on his previous expedition, and was now a Shipwright Officer RN, was on the look out for suitable volunteers.

As I pondered on this casual information, memory of my boyhood's favourite book Nansen's *Farthest North* came vividly back. I remembered how I had longed for similar adventure. I lost no time getting in touch with the officer mentioned, who was then holding an appointment in the Shipwright Officer's drawing office, nearby.

Within the hour I had had an interview and started the ball rolling. As he didn't know me personally he said he would contact the Shipwright Officers under whom I had served and, if everything was satisfactory he would let me know in a day or two. I was not bothered about my professional qualifications, I did not see any difficulty in that direction, but I doubted whether my ten stone four pounds measured up to what I imagined an Antarctic explorer should be.

Two days later the Shipwright Officer came to see me on board *Vanguard*, and told me that from the reports he had already received, I was undoubtedly the man for the job and that he had forwarded my name to Captain Scott with a strong recommendation.

I was now on the tiptoe of expectation. In a few days I received a letter from Captain Scott, informing me that I had been accepted and that application had been made to the Admiralty for approval. As the days passed and there was no reply from the Admiralty, I became unduly anxious, particularly as *Vanguard* was due to sail for Bantry Bay on the west coast of Ireland, to calibrate her guns. I was afraid this might prejudice my chances.

Sailing day arrived and still there was no news.

However, after we had been a few days at Bantry Bay, I received a note by messenger from a friend of mine, a writer in the captain's office, telling me confidentially that my Antarctic job had been approved by the Admiralty and that I was to be discharged forthwith, also that efforts were under way to prevent my leaving the ship before she returned to Devonport.

With this information up my sleeve I went aft to the Captain's office to enquire of the paymaster in charge if there was any news concerning my release by the Admiralty for service with the expedition. He told me there was, and asked if there was any immediate hurry.

To be forewarned is to be forearmed. I said there was certainly need for haste, and pointed out that the expedition was due to sail from London in a months time, and meanwhile there was the refitting the ship, the huts for Winter Quarters, stores and a hundred and one things to be seen to.

I asked him to take me before the Commander, a very keen gunnery expert who was then on Monkey Island (upper bridge) directing calibrating operations. The commander was not easily approachable at the best of times so I was not surprised when the paymaster hesitated to butt in just then. However, he eventually agreed to take me before the Commander and up we climbed to Monkey Island.

It was as I had expected, when the paymaster tried to explain the purpose of my visit the Commander went off the deep end, saying he could not attend to the matter then and in any case there was no boat available to land me. I sensed he was intending to be awkward but I was not to be fobbed off in this manner and told him if there was any difficulty in my getting off the ship I should have to wire Captain Scott. That tore it! He was furious and literally swept us off the bridge.

This little set back did not deter me from making arrangements to leave the ship at short notice, so when later, the Commander sent a message to the effect that he would give me ten minutes to get out of the ship I had time to spare. The boat landed me at Glengariff, not far from where the ship lay, some fourteen miles from the town of Bantry, the terminus of the railway in that direction. It would have been possible to land at Bantry, had a boat been available, but I considered myself fortunate to be landed at all under the circumstances.

My first concern was to obtain transport for myself and three hundred weight of luggage, motor transport was not in general use and practically unknown in this out of the way place. I was fortunate in finding the driver of a jaunting car, who was going to Bantry later in the afternoon to meet some visitors arriving by train.

He said he thought it was a bit of a load for his horse, I thought so too when I saw the horse which appeared to be built on the lines of a greyhound, but under the mellowing influence of a couple of pints of good Irish porter, for which that part of the country was famous, we came to terms.

Whilst waiting I celebrated my good luck so far, and by the time we started for Bantry I was full of the joys of Spring. As the old horse clip-clopped along the hard, dusty road I could see *Vanguard* still engaged on her lawful occasion. At intervals one of her big guns answered the questions with a flash and a roar, then all was peace again as the smoke drifted slowly away and disappeared in the haze. What a picture she made! Britain's latest battleship, - the finest in the world – on that lovely afternoon, riding on the calm waters of the Bay.

In such a setting who would have been bold enough to prophesy that within little more that four years our country would be at war and fighting for very existence, and that during the war I should see the fine ship destroyed in the matter of minutes with most of her gallant crew. I never saw any of my messmates again. After completing a two year commission in the ship, most of them were drafted to HMS *Monmouth*, one of the ships of Admiral Craddock's Squadron sunk by the Germans at the battle of Coronel, 1914. Many years later, when serving in a small scientific research vessel during survey of the Humboldt Current, off the coast of Chile and Peru, the ship was stopped about the position of the battle of Coronel, to pay our respects to the gallant dead.

I arrived at the Royal Naval Barracks, Devonport, on 1 May, and was discharged the following day for service with the British Antarctic Expedition, with instruction to report to Captain Scott, at the offices of the expedition, Victoria Street, London. In spite of the fact that I had volunteered for this job I was escorted to Plymouth by a petty officer who saw me safely on the train, complete with travelling warrant and meal ticket, the Navy never does things by halves.

At Exeter I decided to cash in on my meal ticket at a refreshment buffet on the station. It entitled me, beside sandwiches to a pint of beer. I shall never forget that beer, it was awful just swipes.

On arrival in London I put up at the Union Jack Club in Waterloo Road, which was run exclusively for the services, Navy and Army. The Royal Air Force was not even a dream then, flying being in its infancy. As a matter of fact, Bleriot had recently flown the channel and this was considered a great feat. The club was a boon to servicemen it had all the amenities of a good class hotel with excellent service at modern charges, which I deeply appreciated as I did not know London, having only been there once before on a visit to the White City exhibition.

On my first evening I took a stroll to get my bearings, and coming upon Drury Lane Theatre, where the play 'The Whip' was then running, I took the opportunity of seeing it whilst the going was good and enjoyed it very much.

Francis Davies' shipwright's trunk.

Francis Davies and family, wife Ethel, daughter Beatrice (Maidie), and son Peter.

Captain Robert Falcon Scott CVC RN

Chapter II. Getting Ready to Leave

The following morning I presented myself at the offices of the Expedition for an interview with Captain Scott. I was shown into a waiting-room where there was a great variety of the smaller items of polar equipment – clothing, harness for dogs and ponies, skis, pony snow-shoes, cookers and a hundred and one miscellaneous articles. While I was waiting, another gentleman very bronzed and wearing an ancient raincoat, was shown into the room. He, I noticed was particularly interested in the equipment for the ponies. This was Captain Oates of the Inniskilling Dragoons, although it was not until several days later I knew who he was.

My interview with Captain Scott was very satisfactory from my point of view. He asked me why I wanted to go with the expedition and being satisfied with my replies, went on to explain what would be expected of me. My principal job he said would be the erection of Winter Quarters for the Southern party, which was to make an attempt to reach the South Pole, and the eastern party which was to explore King Edward VII Land at the eastern end of the Great Ice Barrier, discovered by Captain Scott on his first expedition. He also told me that I would be paid £40 a year, adding that if I made a success of the job he wouldn't say what he would do for me, but if on the other hand I failed to come up to scratch I would be for the high jump.

Lieutenant Evans RN the Second in Command of the expedition, was also present at the interview. Antarctic exploration was not new to him, he had served as navigator of *Morning*, when that ship together with *Terra Nova* went to the relief of *Discovery*, beset in the ice of McMurdo Sound on Captain Scott's first expedition. After the interview Lieutenant Evans took me along to the expedition ship then fitting out in the West India Docks.

What I expected to see I don't quite remember, but I was much taken aback when I got my first sight of her, she looked an absolute wreck fit only for the knackers yard, long overdue in fact. A sailing ship was a new experience for me. Up to then all my sailoring had been in ships of the King's Navee where everything was spick and span, regardless of expenses, 'all ship-shape and Bristol fashion' as we say at sea. I certainly saw *Terra Nova* at her worst.

The fact that the yardarms were all askew and the riggers were working aloft added further to the appearance of complete chaos. On the poop shipwrights were extending the saloon to provide additional accommodation for extra personnel and building laboratories for the scientists, amidships a large ice box was being built for transporting frozen meat from New Zealand for the shore parties and, what with all this going on and spare yards and spare rudder hardly a square foot of deck was visible.

Lieutenant Evans had told me going down in the train that he was rather worried about the condition of the ship and that the Board of Trade Surveyor had found so many defects which he wanted made good before he would give a certificate of sea-worthiness that he doubted very much if she could be got ready in time to catch the next Antarctic summer season. There was, however, he said, the possibility that Captain Scott would be elected a member of the Royal Yacht Squadron, and in that case we should no longer be troubled by the surveyor.

Captain Scott wanted his old ship *Discovery* which had been specially built for his first expedition, then owned by the Hudson Bay Company, she was in fact tied up in the West India Dock at the time but they could not be persuaded to part with her, so the next best thing was *Terra Nova*. There were not many of this class of ship to pick and choose from.

After Lieutenant Evans had shown me over the vessel pointing out her many weaknesses, he asked me if I would be afraid to sail in her as she was. It was hardly a fair question to put to me if he wanted a conscientious answer for I would have sailed in anything for the privilege of going on such an adventure.

Battered and scarred as she was, she still remained a fine ship having been soundly and truly built of well seasoned timber some twenty five years earlier. On one or two occasions she had been badly squeezed in the pack-ice, once so badly, I was told, that all her hatches were out of shape. For many seasons she had been sailing out of St. Johns, Newfoundland, and afterwards was badly neglected, possibly due to circumstances over which her owners had no control, for there was not much profit in the whaling industry in those days in spite of the dangers and hardships inseparable from that calling. She was also in a very filthy condition though in that respect, no worse than other ships engaged in this unpleasant occupation.

Under four hundred tons register she was built at Dundee in 1885, barque rigged with auxiliary steam power for pushing through the ice. Originally she was fitted with a two bladed propeller that was hoisted in over the trunk when she was under sail only, reminiscent of the days of sail and steam in the Royal Navy when up funnel and down screw was a familiar pipe. Sometime since she had been fitted with four bladed propellers and this made her a bad sailor, it dragged like a sea anchor. She had the most beautiful hull form I have ever seen and from the point of view of stability she was very seaworthy in spite of her heavy top hamper of masts and yards, a grand old lady of the sea.

Whilst getting my first once over of the ship, I was introduced to three of my new shipmates, *[note 1] Mick Crean, Taff Evans and Bill Smythe who were busy with the riggers sending up the yards. All had served in *Discovery* with Captain Scott on his first expedition and all had been seamen petty officers in the Royal Navy, Taff Evans and Mick Crean were still serving and had been released by the Admiralty for service with the expedition. Bill Smythe had found the Navy not to his liking after life in the Antarctic and had taken his discharge at the end of his first period of twelve years and had since sailed in tramps, he had now shipped as sailmaker.

They were all very fine seamen, such as we are not likely to see again, particularly as the sailing ship has almost had its day. What characters they were, Mick, with his ever ready smile and Irish wit was for ever chewing 'baccy, the quid rarely left his cheek except perhaps on odd occasions when he might be sent for by an officer, then it was transferred to his cap. His scalp, with its thinning, unkempt hair to which fragments of tobacco clung like tea leaves was stained brown by the juice of the quids which had found a temporary resting place.

Taff was enormously powerful and might have been a model for the man on the posters advertising a well known brand of stout. He had on more than one occasion been one of the field gun's crew representing the Portsmouth Command at the Royal Tournament, Olympia. At dinner time these three worthies suggested they should show me where to get some 'chow'. From the knowing look on their faces I guessed they were sizing me up, wondering if I would be good for a pint or two of 'Harry Freemans' (beer). I was really very pleased to be included in their company as I felt quite lost on the beach on my own and it would, I thought, be a good opportunity to pay my footing, which according to ancient nautical custom I should be expected to do sooner than later.

We went to a pub just outside the dock gates, called oddly enough 'The North Pole', as we passed the policemen on duty at the dock gates he handed Mick a can which Mick accepted without a word being spoken on either side. I was to find out the meaning of the can in the days that followed to my cost, custom demanded a pint for the policeman on the way back.

Knowing chaps those 'bobbies'. Never did any of the crew slip out to the pub for a quiet one without being handed the can, what they did with it all puzzled me, they must have hollow legs.

'The North Pole' became a home from home for most of the crew till the ship sailed. For threepence we could get enough bread and cheese with pickled onions for a good tuck in, and as much beer as we could pay for or strap at tuppence a pint from 6 p.m. till midnight, with civility thrown in for good measure.

That was in the 'bad old days' of course, before this fine old country of ours had shipped the bonds of freedom, when we really were free and it was not considered necessary to pipe the fact every time the bell struck to bring it to notice. It is pleasant to remember having lived in those days. Times have changed a good deal since those not so far off times. I'm afraid the common man has swallowed the bait of democracy hook, line and sinker, and sprung the trap, and it will be a long time before he will gnaw his way to freedom again. As for my own generation. Well, we've had it, Chums!

A day or two after I joined the ship she was inclined for stability by a ship constructor from the Admiralty. Amongst the people who were assisting the seamen to hump the pig iron ballast from side to side of the ship to incline her, was Captain Oates, rigged in a serge suit and peak cap. None of us knew who he was up to this time and many were the guesses, all wrong.

I joined the ship in civvies, wearing a bowler hat, then very fashionable. Mick, when introduced mistook me for one of the scientists and said 'pleased to meet you, Sir' he never got over that mistake on his part and often ragged me about it, saying, 'me, calling a …"hard hat", Sir!'

The seamen generally worked aloft in bare feet, even when fitting out. Taff and Mick were in digs together, Taff used to spin the yarn that his first job mornings was to separate Mick from the sheets which had become stuck to the tar on his feet.

On one occasion they went shopping somewhere in Petticoat Lane and York Road. Taff wanted a 'civvy' suit and naturally did not want to pay too much for it, more clothes meant less beer. He spotted a suit on one of the barrows and thought it just what he was looking for. He tried the coat on as the barrow man, gathering a fathom in his hand at the back, told him it fitted him like ' de paper on de vall.' The salesman wrapped it in paper and off they sailed, very pleased with themselves. It was rather late that night when they reached their digs, even so, Taff wanted to admire his new suit. When he opened the parcel all he found were two potato sacks. Either the barrow man, or somebody had rung the changes on him in one of the many pubs they had visited.

My most important job before the ship sailed was to see that the huts to be erected in the Antarctic as Winter Quarters for the shore parties were complete in every detail. Huts they styled, but they were more like Parish halls. The larger of the two for the Southern party was fifty by twenty five feet and the other for the Eastern party twenty five by twenty five feet. They were not made in sections as this would take up too much storage in the ship but the frames were mortised and tenoned as far as possible for easy erection. The framework was being made and erected, temporarily, on some waste ground at Poplar. This I could see was all in order, but the timber for cladding the huts was being supplied from the sawmills direct to the ship and this was not so satisfactory. I had asked the firm's representative, a rather garrulous old gentleman on several occasions to let me have a copy of the orders so that I could check on it, but he always put me off with the assurance that there was plenty, and to spare of boarding, and invariably added that he had erected this class of building in every country in Europe.

I did not intend, however, to take any chances and eventually went to the sawmills myself and got a copy of the order. On checking it I found it was being supplied in squares (100 superficial feet) irrespective of the precise length of boards required. This meant a loss of at least 10% in cutting, just waste. I spoke first to the foreman on the job about the shortage and I could see he more or less agreed, but I suppose it was more than his job was worth to tell me so. Then I tackled the representative when next he came on the job. He still tried the old assurance subterfuge but this time it wasn't going to work, and as I could not persuade him to do anything about it and the ship was due to sail in a few days, I reported the matter to Captain Scott.

The following day I received a wire from Captain Scott to appear at the office. There I found the garrulous old gentleman and his foreman. Captain Scott went into the matter very carefully and at first told me he was quite satisfied there was sufficient boarding. I again pointed out the amount of waste in cutting, due to the boards not being supplied in proper lengths and said I could not take any responsibility for erecting the huts unless I had the material I required. Captain Scott then saw what I was driving at and told the old gentleman that unless he was able to satisfy me there was sufficient material, before the ship sailed, the firm would not get a penny piece for the job.

I got what was necessary and apologies from the old chap, who admitted he had made a mistake not a serious one where a few hundred feet was easily obtainable, but rather so, in the Antarctic. He came down to see me before we sailed and tried to slip me half a sovereign, but I told him I wasn't that sort of chap. He then promised to send me a box of cigars for Christmas perhaps he did, anyhow I never received them. During a conversation towards the end of the expedition, Lieutenant Pennell said to me, 'You have done some fine jobs from time to time, Davies, which do you consider your best?' I replied, 'getting the huts away complete from London.'

Up to within a day or two of the ship sailing I felt the whole show as far as I was concerned might come unstuck, and this caused me great anxiety for up to that time I had not been medically examined by the expedition doctors and I really did not fancy my chances. However, my fears were groundless, I managed to get by all right.

Before we departed some members of a learned society calipered our heads and took the colour of our eyes and skin, presumably to note if these were affected in any way by the intense cold. They were elderly people and as we did not see them when we returned I suppose they had passed on to where these matters are of little concern.

All the stores were collected in a warehouse close to where the ship lay, and were sorted out for loading in their proper order. These had in many cases been supplied by the firms who manufactured them, free of charge, and were packed in light three ply Venesta cases, each about fifty pounds in weight for easy handling, the contents being stencilled on the outside with a distinctive band either red, green or black to indicate whether the case was for the Southern Party, Eastern Party or ship's party.

We were all working like beavers to get the ship ready by the appointed date, but it was not all work and no play, our evenings were generally free and we made the best of them seeing the sights of London.

Being sailors it is hardly necessary to mention that the famous taverns came in for more than a fair share of attention. Beer was beer in those days, unlike the present "near water" infliction of today, and at a price that even the common folk could afford.

In this atmosphere the crew got to know each other and happy memories of those hilarious evenings were often recalled during the hard years that followed.

At last sailing day, 1 June 1910, hove round. Captain Scott had been elected a member of the Royal Yacht Squadron and the hoisting of the white Ensign for the first time was made the occasion of a little ceremony on the poop, Lady Bridgeman, wife of Admiral Sir Francis Bridgeman, a Lord of the Admiralty, breaking the flag at the peak.

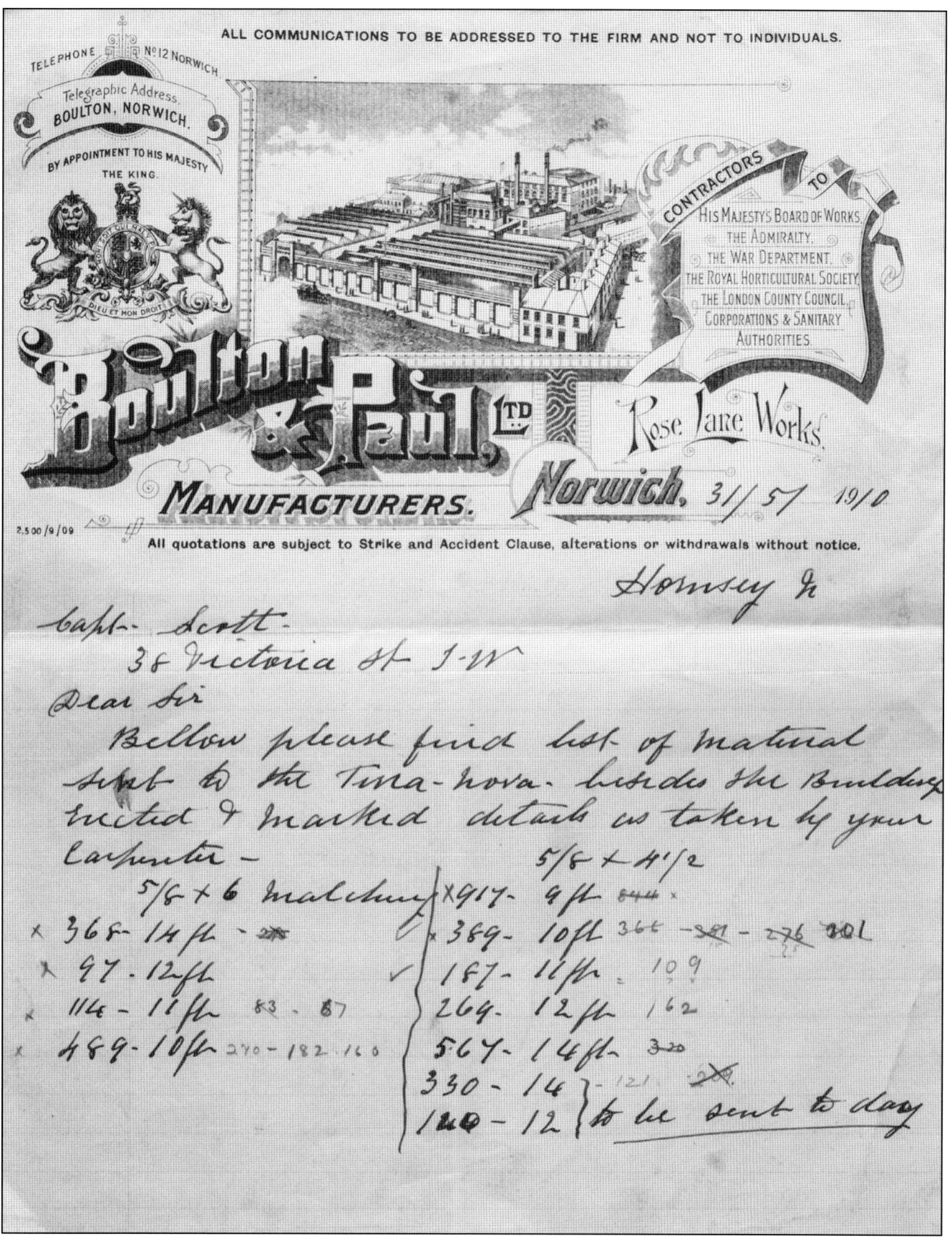

Copy of list of materials for construction of huts.
Davies argued with suppliers and was adamant about ensuring sufficient quantities were acquired

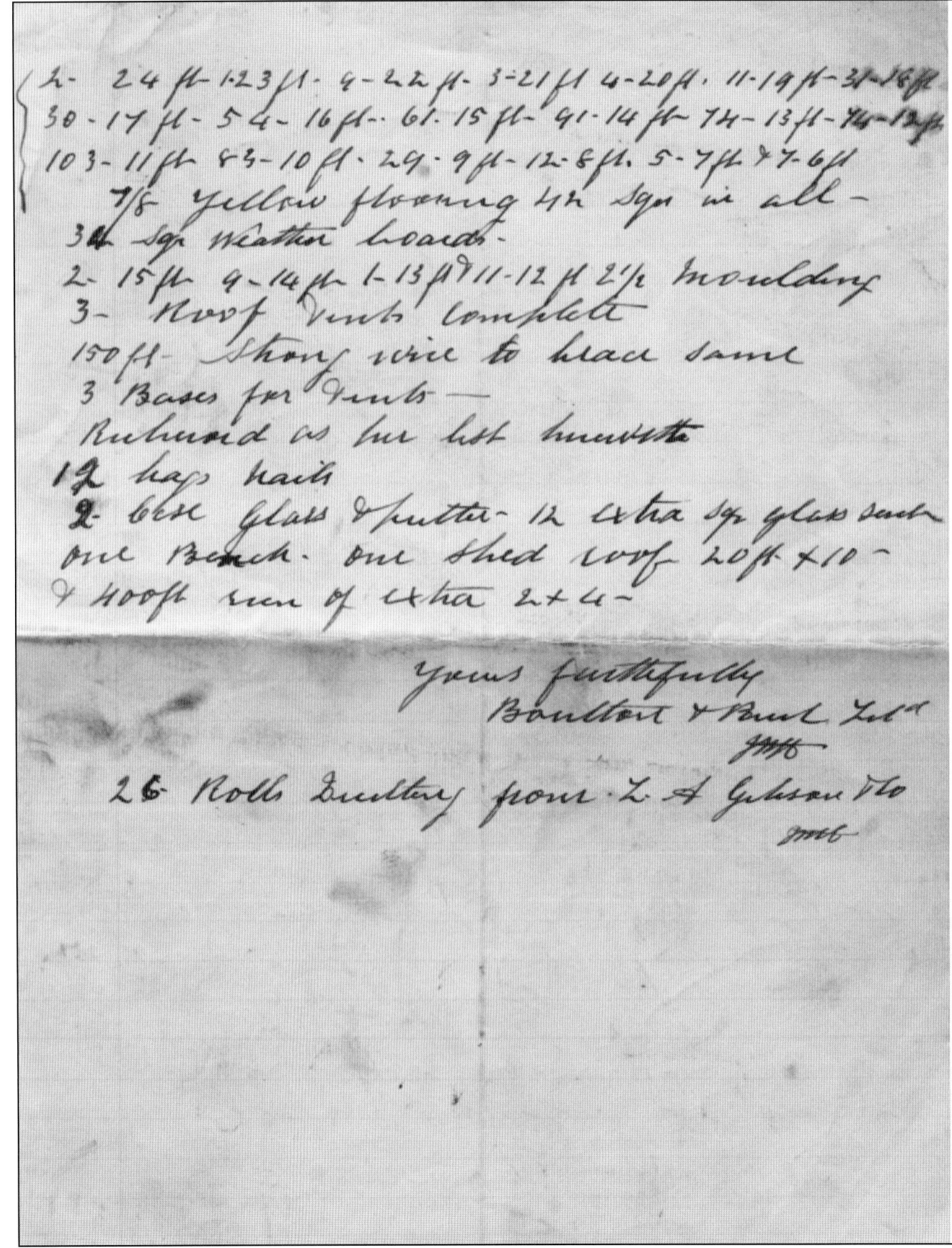

Many of the stores were still on deck, there had been such a rush in the last hour or two to get everything on board. On the day of departure there was a very big crowd to bid the farewell and God speed, also much cheering. As the ship cast off from the jetty the crowd struck up 'All The Nice Girls Love A Sailor'. The girls from 'The North Pole' were there too, right in front shedding tears as large as gooseberries. Many of the ships in port were dressed over all (flying strings of flags) and blew their sirens as we passed down the river.

Our first port of call was Portsmouth to pick up some special Navigational instruments which were being loaned by the Admiralty. Leaving Portsmouth, on our way down Channel en route for Cardiff, we steamed into Portland harbour and through the fleet which was assembling for a review at Spithead in honour of a visit from the King of Spain. It was a grand sight, Britain's sure shield, the Royal Navy. As we passed through the lines of ships their crews cheered us on our way, it was a wonderful send off.

At Cardiff we loaded thirty tons of coal briquettes, fuel for the shore parties, and about four hundred tons of coal, gifts from the shipowners of Cardiff. We were entertained right royally at Cardiff. I remember one rather amusing incident which happened at a reception given in honour of Captain Scott by the Lord Mayor at the City Hall, to which the ship's company was invited. As one of the seamen, accompanied by a lady friend was ascending the marble stairs behind a lady wearing a train, he accidentally stepped on it.

The lady turned round and in very dignified and haughty manner, fixed him with her lorgnette. He was so caught aback he almost forgot to apologise.

Cardiff also did the expedition proud in the matter of funds. One night at a dinner given by the shipowners to which we were all invited, Mr. Dan Radcliff, the shipowner, gave a fairly large sum and ragged his brother, Henry, to follow suit.

Things got rather hectic towards the end of the evening, most of us were well oiled when Mr. Dan, in a burst of exuberance got on the table and walked from end to end. Dodging the bottles just to show us he was still able, I suppose. We were also entertained at the Stock Exchange to lunch and the hospitality of its members wanted some standing up to.

Captain Scott, to show his appreciation of the exceptionally practical help and interest in the expedition by the citizens of Cardiff, promised the Lord Mayor that Cardiff should be the port of call in the United Kingdom on his return – three years later to the date on which we sailed *Terra Nova* returned to Cardiff - ALAS Without Captain Scott and his gallant companions who had perished with him.

After the ship had been loaded with coal a very serious leak developed, making nearly four feet of water a watch. It was thought to be somewhere in the bows, anyhow a certain amount could be heard trickling in near the stern. However, as it could not be located by the shipwrights it was decided to proceed and hope for the best, and to see what could be done in Lyttleton, where she would be dry docked before leaving for the Antarctic. This leek caused us a great deal of trouble from first to last, before it was discovered quite by accident nearly two years later.

Captain Scott did not sail in the ship from Cardiff, joining her later at Simonstown, South Africa. He had a very full programme of lectures to raise badly needed funds for the venture.

Mrs Scott on board Terra Nova just before the ship sailed from England with Lieut Henry R. Bowers and Captain Lawrence E.G. Oates

Lady Bridgeman - Raising Ensign

Petty Officer Edger 'Taff' Evans

Chapter III. Setting Sail

Lieutenant Evans was in command of the ship when she sailed on 15 June, 1910. The other members of the ship's company were Lieutenant Campbell, RN Chief Officer (Later leader of the party landed at Cape Adare) Lieutenant Pennell, Navigating Officer – lost in HMS *Queen Mary* at the Battle of Jutland, 1916. – Lieutenant Rennick, RN, Second Officer, and hydrographic Surveyor – also lost during the first world war in HMS *Hogue* when she was torpedoed by a German submarine, U29, 1914; Lieutenant Bowers RIM, Third officer, and Sailing Master – perished with Captain Scott; Captain Oates, Inniskilling Dragoons, who later had charge of the ponies who also perished on the return journey from the Pole; Surgeon Lieutenant Commander Murray Levick RN, ship's doctor – landed with Cape Adare party. Surgeon Lieutenant Atkinson RN, an Engineer Officer RN, who did not get further than Melbourne, outward bound; Lieutenant Gran of the Royal Norwegian Navy, ski Expert, Doctor Wilson, Chief of Scientific Staff perished with Captain Scott; Doctor Simpson and Mr Wright, physicists, Doctor Nelson and Doctor Lillie, biologists and Mr Cherry Gerrard, zoologist made up the Afterguard.

Amongst the crew under the foc'le were a Bos'n RN who left the ship at Lyttleton, outward bound. Alf Cheetham, merchant Navy, Bos'n of the ship; Billy Williams CERA, RN, Second Engineer; Bill Lashley, Chief Stoker RN, Third Engineer; Taff Evans and Mick Crean, Petty Officers RN, in charge of Port and Starboard watches respectively; Bill Heald and Tom Williamson – served in *Discovery*; Patsy Keohane, Joe Ford, Fred Parsons, Arthur Browning – Harry Dickason, all Petty Officers in the Royal Navy; Hugh Mather, Petty Officer, RNVR, Walter Archer, Chief Steward and Chef; Tom Clissold, assistant Chef; Lofty Hooper and Bill Neald, Stewards; Bob Brissenden , Stoker Petty Officer RN – drowned during a survey of the French Pass, New Zealand,1912 – Tom Mckensie, Billy Burton, firemen also Stoker Petty Officers RN, Bill Smythe, Sailmaker, and myself, Carpenter. There were a few seamen in addition to those mentioned by name but these mostly faded out before the ship left Lyttleton, on her first voyage South.

We had two dogs on board presented by Peary, first to reach the North Pole, called 'Peary' and 'Cook' by the ship's company. Cook claimed to have reached the North Pole before Peary the same year, but was later discredited. The name Cook was officially changed to 'Yank' for obvious reasons.

In spite of the fact that it was generally acknowledged this was the best equipped expedition that had ever left the shores of this country, or any other country for that matter for Polar exploration, most of the ship's fittings were antiquated and we were minus radio. Maybe it will not be out of place to describe some of these ancient appliances which were more or less my special 'babies' and took an awful lot of nursing.

Pride of place must, I think, be given to the windless. This piece of 'Armstrong's patent' (hand) machinery 'lived' just under the break of the foc'le and was of course used for heaving up the anchor and operated by the Capstan on the foc'le which was hove round by means of wooden bars (capstan bars). The motive power being supplied by every man jack of the crew who could be clapped on the bars to the tune of a favourite shanty, generally 'Rio Grande'. When it first came under my notice it looked almost like a solid billet of rusted iron though the working parts could hardly be described as delicate. I should think somebody had at one time had a brain wave and fitted a gypsy (endless chain) between the fore winch and the windless, so that it could be operated by steam instead of by hand.

Terra Nova

In order to connect the winch and the windless the gypsy had to be passed through the galley, which had two small shutters fitted, one on the fore end and one on the after end for this purpose.

To prepare for heaving up the anchor, the gypsy – the links of which fitted in snugs on the sprocket wheels on the winch and windless – was passed over the two sprockets and the two ends brought together and connected by riveting two half links to form an endless chain. All was then ready for heaving in the cable. I took my place at the windless ready to tend the brake and an engineer stood by the steam valve to work the winch. From my position I could neither see the officer giving orders, or the engineer, nor could the engineer see the officer. At the order 'heave in', I eased the brake and the engineer turned on the steam. With a jig and jerk the strain came on the gypsy and slowly the cable started to come in. In a few seconds the gypsy would have stretched to such an extent it flapped up and down in the galley, sending the cooking utensils flying in every direction to the accompaniment of two beautiful flows of nautical language from the chef and his mate.

There was such an awful din it was impossible to hear any orders passed, these were conveyed in a sort of 'tic-tac' code by seamen posted at strategic points. Often it would be necessary to stop and shorten the gypsy by cutting off two or three links and reconnecting with half links. Heath Robinson had nothing on this, but all hands thought it a wonderful idea as it saved them hours of 'roundey-come-roundey' on the capstan.

The next in order of seniority and perhaps most important of all, was the hand pump. This functioned on deck between the main rigging. It was a heavy cast iron affair and had four vertical plungers and was really two independent pumps for there were two separate suction pipes from the bilges. The water was discharged on deck through the wide mouth of the Old Man of the Sea, whose head adorned the sunny side of the pump.

Under good conditions it was a most efficient type of pump, but with dirty bilges such as these were it was another story. It gave me endless trouble when the weather was bad. I almost had to sleep beside it. Unfortunately the bilges under the ceiling (inner bottom planking) were not get-at-able for cleaning and the muck from them found its way under the leather seated valves, putting the pump completely out of action.

This happened very often during bad weather, making it necessary to completely strip the pump in order to clean the valves which meant removing the plungers to get at the foot valves. These were portable and fitted on heavy tapered metal plugs, wrapped round with ordinary lamp wick to pack them tightly in their places. On top of each plug was a large eye for lifting them out. If these were driven tightly I had difficulty in getting them out, if they were not tight enough they worked loose on their own and put the pump out of action.

The steering gear was the least troublesome of my 'babes'. It was right aft on the poop and from here the helmsman steered the ship. There was no shelter for him in any shape or form, and in bad weather and particularly on dark nights with a roaring, towering, following sea which threatened to engulf the ship every few seconds, it was really frightening. The only spot of comfort was the dim light from the binnacle lamp of the steering compass, no human being nearer than the bridge. It was of the most simple design, just a chain, one end of which was attached to the tiller and the other end rove through a block on the deck at the ship's side, then three or four turns over the barrel of the wheel through another block on the opposite side of the ship and finally secured to the tiller.

The steering wheel was nearly six feet in diameter. There were, two wheels, one at the fore end and one at the after end of the barrel. It had one great advantage, the helm could be put over quickly, but unlike screw steering gear it was not self holding and had to be held against the sea constantly by the helmsman, who received every shock from the heavy seas when they struck the rudder. It was not unknown for the helmsman to be thrown right over the wheel when the rudder was hit by a particularly heavy sea.

To take a certain amount of the strain, two rope lanyards were fitted one on either side of the wheel attached to heavy ringbolts in the deck. With one of these the helmsman took a turn round the wheel and held it in hand so that it could be released instantly if necessary. At times there was an additional helmsman, a lee helmsman, particularly when it was necessary to use the rudder a lot for instance when working through pack-ice, when a two hour trick at the wheel was no job for a weakling. The Officer of the watch always kept half an eye on the helmsman, whose face was visible even on the darkest night, in the dim light of the binnacle as he watched the compass.

The crew were always very considerate to one another and never forgot to take along a steaming hot cup of cocoa at least once during the trick. How much one appreciates the simple and commonplace things of life under conditions of great emotional strain.

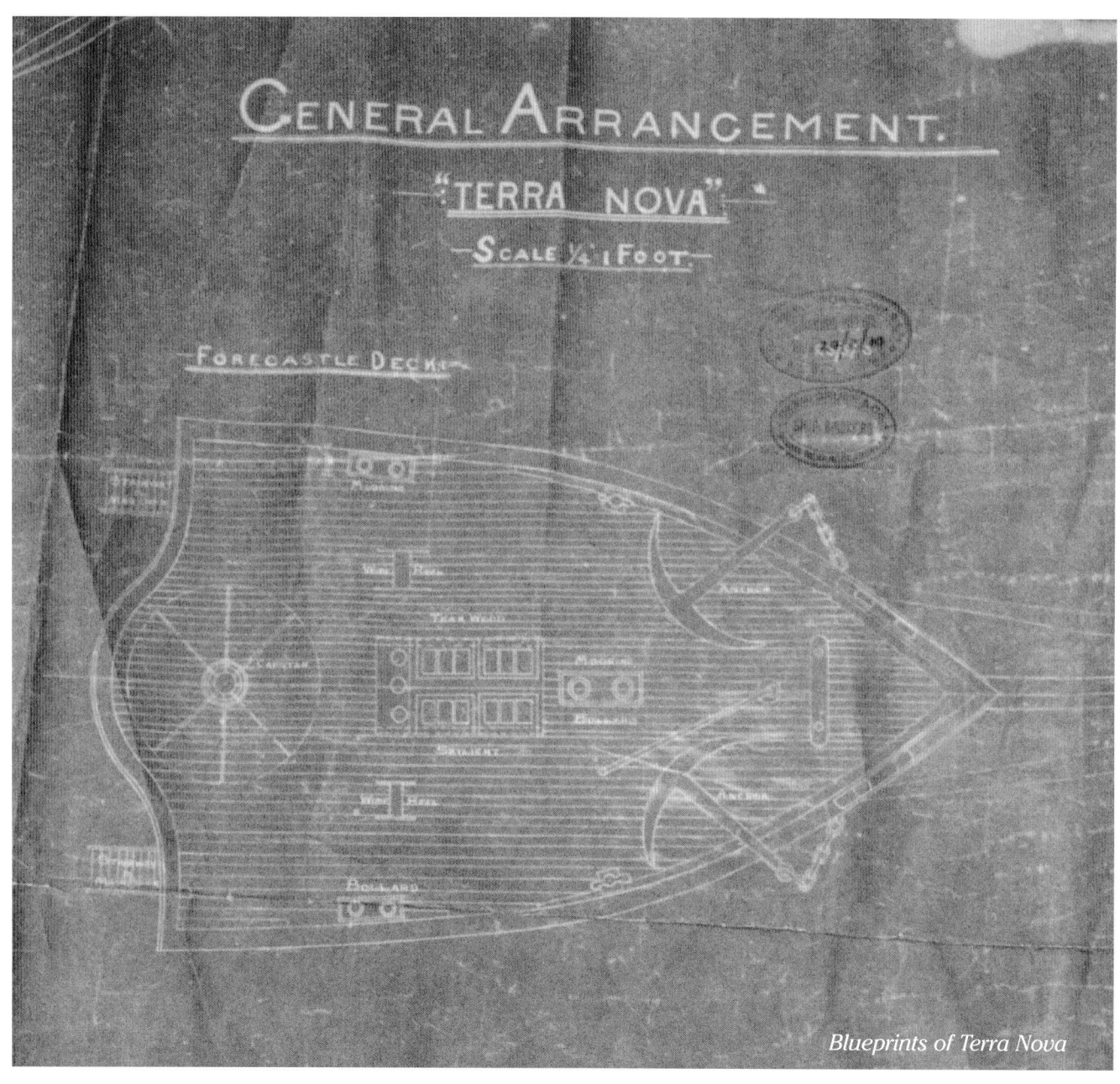

Blueprints of Terra Nova

Chapter IV. Heading South

As soon as the ship cleared land her anchors were got on board and securely lashed on the foc'le and the cables unbent and stowed in the chain-lockers. The hawse pipes, which were open to the mess deck under the foc'le, were then plugged with large wooden plugs (hawse plugs) and well cemented before fitting an iron plate over the whole inside, which in turn was also cemented. This kept the mess deck dry and snug. The cable would not be again bent until we approached our next port of call, which might be a matter of weeks or even months.

After leaving Cardiff we made for Funchal, Madeira. The weather was very fine with light head winds so we steamed all the way. This gave an opportunity to get all the loose stores that had been brought aboard at the last moment properly stowed and everything as it should be aloft.

We had been at sea a few days when it was reported to me that the skid beams over the galley, which had been fitted in London for the stowage of two of the lifeboats, were 'on a wind' as the sailors put it, meaning they were working badly with the rolling of the ship. On examination I discovered the reason, both boats had been loaded to capacity with all the crockery, glassware, cases of bottled stores and other fragile articles for safety. There must have been at least a ton or so. The skids were never intended, or expected to stand up to such an undue strain. I fixed them by fitting two heavy struts, or shores, one on each side of the ship. These were quite effective and were still in place when we left the ship at the end of the expedition.

From the very outset the spirit of the expedition was the spirit of Drake. The gentlemen pulled with the mariners, and the success achieved was in a great measure due to the splendid cooperation of all hands irrespective of rank. I do not suggest for one moment there was any easy going familiarity, such as one meets everywhere today, for we had the greatest respect and admiration for our officers and would have followed them anywhere. They were of course 'hand picked' one might say, there were no less than eleven thousand applications from people in all walks of life, eager to sail this great adventure.

All the Afterguard, with the exception of the Captain and the watch keeping officers, were detailed to work with the seamen in watches, making and furling sail, and even the scientists soon became very much at home whether aloft or sweating up the braces on deck.

Captain Oates and Doctor Atkinson were also told to work with me as carpenters mates when not otherwise employed on deck. Captain Oates was extremely keen and soon became very useful. I often gave him jobs which he did entirely on his own.

Between times the scientists fitted in the work of their own particular ologies. One of them, Doctor Nelson, kept regular watches in the engine room as engineer in charge of the watch when the ship was under steam. Often, when passing through the tropics, an officer would take over the fireman's watch just to give him a spell, and at the same time earn a little fresh water from the hot well for a bath.

When the ship was steaming, coal had to be trimmed from the main hold to the coal bunkers on the poop. This was another job which the Afterguard made their own during the forenoons.

They also did most of the hand pumping when the ship was under sail only. Everybody was kept fully occupied and there was never any lack of volunteers for any old job that turned up.

The food was of the very best quality and plenty of it, full and plenty, no waste as it says in ship articles. We had two excellent chefs who took a pride in their craft and only once during the whole three years did we miss a meal, a breakfast, during exceptionally bad weather and even then we had a large slab of chocolate apiece.

About eight days after leaving Cardiff we arrived at Madeira. What a beautiful sight the island was from the harbour in the sunshine of a lovely summer's afternoon. Almost as soon as we had dropped anchor we were surrounded by scores of bum boats loaded with all kinds of wares for sale, or even exchange for old clothes, fruit, cigars, perfume, jewellery, a great variety of table and household linen, gorgeous ladies' underwear, most of it the beautiful needle work for which the island is well known.

The guides, all clamouring at once to be engaged to show off the sights of the island, were a scourge. The best way it seemed to rid oneself of this nuisance was to engage the most pugnacious member of the fraternity to keep the remainder off. After dark the place was like a scene from fairyland, illuminated with myriads of tiny coloured lamps amongst the trees which with the strumming of guitars, the scents of luscious fruits and flowers on the balmy air was something not easily forgotten.

A voluntary boat's crew from the Afterguard, including the Captain, came in for us in the evening, I'm afraid I, at least, did not feel very comfortable to be sitting back whilst the officers bent their backs to the oars. It was something I had not been used to in 'Andrew Miller' (RN) and just goes to show how deep are the habits formed by long training.

Before sailing we took a good supply of fresh fruit and vegetables aboard. Amongst the fruit was a large basket of lovely ripe apricots, I had never seen them before, except out of tins, and didn't even guess what they were.

The weather continued fine as we steamed southwards to pick up the north east Trade Winds. Shoals of flying fish continually bobbed out and in the long, oily swell, some no bigger than a good size bumble bee whilst the largest of them were a little larger than a herring. They shoot out of the water to escape from their enemies, the porpoises and dolphins, and can fly comparatively long distances. It is always a point for discussion as to whether they fly or just volplane. I have watched them very carefully but cannot claim to have seen their wings move in the same way a bird flies. It was particularly noticeable they kept going by just touching the water with the lower half of their tails, which they wriggled violently and gained sufficient impetus to continue their flight. I think a hundred to two hundred yards is easily within their power. At night, after dark, it is usual to fix a light on the rail of the ship to attract these fish, and it is very rare not to find a few on deck in the morning. They are very good eating and well worth no little trouble to get a nice fresh fish breakfast. They have very large wings, almost transparent, and beautifully marked like butterflies. The sailors used to set these wings on pieces of board to dry before varnishing them as souvenirs for their friends at home. It was generally the case of many were dried but few were varnished. The ship's cat soon found out where they came from and would wait in place of vantage for them to fly on board.

Soon we picked up the Trade Winds which blow steadily from the north east. Fires were drawn to save coal for the times when we would get head winds, or perhaps no winds at all, as in the Doldrums and we were then under sail only.

So far the voyage had been almost like a yachting cruise and I was enjoying life to the full. The weather got warmer as we worked our way through the Tropics and became very stuffy below decks. Many of us slept on the foc'le. It was lovely to lie and gaze aloft at the sails billowing to the fresh breeze in the moonlight, the only sounds being the slight sighing of the wind through the rigging, the swish, swish of the waters round the bows as the ship sped onward, an occasional shout from the lookout as he struck the bell and hailed the bridge to report 'all's well'.

Now that we were under sail some of the ship's weaknesses became apparent and I, with my mates, was kept very busy. The hand pump too took up a lot of my time as we now depended on it entirely to free the bilges of water from the leak.

One day the main top-gallant yard stripped the lightning conductor from the top gallant mast. This was a number one priority job. It was not the easiest of places to work a breast drill, to bore holes in the copper strips to rivet the ends of the conductor together. The rolling of the ship caused the breaking of more than one drill before the job was completed. Captain Oates was my mate at the time. As soon as this job was done, the out riggers to the main top-gallant mast back stays carried away and had to be renewed.

About this time the after end of the coal-bunker came away from the ship's side entirely, the fire and bilge pump with it. With the assistance of the Norgwegian Lieutenant. I made three large angle irons and plate brackets, using a rivet forge in the stoke-hold. This made a good strong job and never showed any sign of weakness afterwards. Before I had completed the bunker job I was called on deck. The hoop on the heel of the spanker boom had come off and had, of course, to be replaced at once, so that the spanker could again be set.

Often during the night I would have to strip the hand pump to clean valves. I didn't mind this job as I generally got a good stiff tot of Peter Dawsons whisky from the officer of the watch. This I appreciated very much for more reasons than one, for I missed my daily tot of rum that I had been used to in the Royal Navy.

'Splicing the Main Brace', a tot for all hands was reserved for special occasions, such as birthdays or when the weather was extra bad and both watches were on deck, or any other event or occasion that might serve as an excuse for a celebration. Captain Oates thought the anniversary of Napoleon's birthday was an event of sufficient importance for celebration. I don't know whether it actually was the 'Little corporal's' birthday or not, and I'm not quite sure that it worked, but at least it was worth a try on.

When my birthday hove round I went aft, full of the joys of anticipation to collect the usual two bottles to celebrate the occasion with all hands for'd. I got a shock when I was told I had already celebrated my birthday at least four times and the ship had only been in commission about eight months! I chose to believe it was a bit of a leg pull, surely I was not so absent minded as that or was I?

It generally seemed to blow gale force on Sundays, and church which was usually held in the saloon would have to be postponed. Our first real gale happened on a Sabbath. The ship was struck by a sudden squall that split the main top-gallant sail from head to foot before the halyards could be let go to lower the yard.

The top-gallant mast bent like a whip stick. I was sent aloft to examine the mast as it was thought it might have sprung. Whilst I was giving it the once over, standing on the main upper top sail yard and hanging on to the rigging, the ship was stuck by another squall and the halyard of the yard, on which I was standing, was let go with a run. I managed to jump into the rigging and got off with only the loss of my cap which sailed off to leeward it might have been worse.

All hands had a pretty tough time that day hoisting up the yard and then having to let them go again and again whilst the seamen unbent the split sail and bent a new one. The Captain thought this an occasion that warranted the 'Splicing of the Main Brace', even a gale might have a silver lining.

By this time we were all becoming real 'shellbacks', almost buccaneers, at least that's how we felt as we hauled on the ropes clad only in a pair of dungaree trousers. The life was certainly agreeing with me for I had already put on a stone and felt fitter than ever before. We took full advantage of the heavy, tropical rain to do our dhobeying (washing clothes). Scuppers were plugged to accumulate the rain water on deck. Then in our birthday suits, we had combined shower bath and washing day. When there was little wind and the ship was scarcely making any headway, some of the officers took their bath over the side on the end of the rope, but an ominous shadow under the ship (Johnny shark) on one occasion caused this style of bathing to lose its popularity. It was considered wiser to take the 'waters' in smaller doses by means of a bucket dipped over the side.

Going through the hottest part of the Tropics we were each allowed two small bottles of a very light beer daily. This was greatly appreciated as a change from the ship's water which however carefully the tanks were looked after, became flat and discoloured with the rolling of the ship. Occasionally a bottle of stout, of a well known export brand, was substitute for the beer. This was excellent but it had one disadvantage, it was odds on the entire contents of the bottle would shoot up under the deckhead and of course be lost.

The ship's cat in a bunk of its own was comfortable whatever the weather conditions were like for rest of the crew

Commander Evans

Chapter V. Crossing the Line

As we neared the Doldrums in the vicinity of the Equator the wind fell light and the ship rolled tremendously as the sails flapped against the mast. It became a saying in the foc'le, 'moderate roll rings the bell (ship's bell) great roll brings out the cooks', this was, in fact quite true.

Mr Mate asked me if I thought I could fit a grating or something on the deck of the galley to give the cooks a better foothold. I made the grating and it met with the unstinted approval of the cooks until one day when the ship gave a tremendous roll and the bell, true to form went ding, and the two cooks shot out of the galley, one on all fours and the other on his back into the scuppers, followed by tins of tripe and onions that had just been opened for the Afterguard's lunch.

The crew were all at lunch at the time and a great cheer went up as the cooks sorted themselves out and started to scrape off the tripe and onions a lot of which had found its way into their sea boots. Such a flow of nautical language was surely never heard before, what those rude men called me and my handiwork is nobody's business. The gratings were voted unsuitable, the cooks could not get their feet free quickly enough to save themselves. After that they stuck to the old sailing ship dodge of laying wet sacks on the deck which was found to answer best of all. Incidents such as these caused much amusement fore and aft, but left no rancour.

One day Captain Oates and Alf, the Bos'n were painting my old friend the windlass. I was working away at my bench, which lived close to the windlass. Alf was spinning Captain Oates a bender about one Jarcomo Guiseppe, or that's what the name sounded like to me, whom I gathered Alf had been shipmates with a long time before. It was one of those once upon a time stories of loves young dream. I had one ear cocked, listening, as I got on with my work. Just as Alf got to the most exiting part of the story. Cookie, who had evidently been listening also, emerged from his galley and chanted. 'A cow climbed up an apple tree', 'Oh! You liar'.

At Cardiff, Brownie had brought a tiny black kitten aboard. The cat became a great favourite with all and was looked upon as the ship's mascot. Brownie made him a tiny hammock. Complete in every detail, including a blanket which was along under the foc'le close to the heel of the jibboom. He grew into a fine cat and was a proper old 'shellback'. I'm quite sure he used to dodge the column in bad weather by refusing to turn out of his hammock. Nothing ever seamed to disturb him, he was quite blaze.

Having no wireless, we lived in a world of our own. When we left port we disappeared into the 'Blue' and as we followed the sailing ships tracks and were off the usual shipping lanes we rarely saw another ship for weeks, or even months until we neared the next port of call. On one occasion we passed a full rigged ship carrying everything she had. She was of French nationality; a very beautiful sight with her sails billowing out to a good sailing breeze.

The scientists were busy on their own jobs whenever possible. The biologists at regular intervals towed very fine silk nets astern to catch samples of plankton in order to get a picture of its distribution in the sea. When the nets were out the heads (WCs) were fastened and the word passed that nothing was to be thrown overboard until they were hauled. Sometimes, perhaps by force of habit or necessity, these orders misfired and when the nets were hauled up would possibly contain an empty tin or two from the galley or even something more unpleasant.

On these occasions the seamen were sure to take more than a passing interest. One of them would approach the scientist and, in a guileless manner ask him if it was a good catch. He would explain to them that unfortunately a couple of empty tins and some excreta had got into the nets. With a face as expressionless as a mask they would ask him if this was some new, or rare specimen and were not satisfied until he had told them what it was in plain, if somewhat vulgar language. This answer would fill them with glee and they would hasten for'd to tell their messmates who, one by one, would find some job on the poop where the scientist was sorting the catch, and ask the same question in order to get him to repeat his explanation. The scientist fell for it every time, he was probably too nice to notice their apparent ignorance.

When we reached the Doldrums, the area of calms, steam was raised and the sails furled until we picked up the South East Trades on the other side of the Equator.

There was much excitement as we neared the Equator over the crossing of the Line ceremony. Preparations commenced days before, suitable costumes etc. for Neptune and the members of his Court were made from all sorts of oddments. The 'Royal' robe for Neptune's Consort, the fair Amphritrite, was a creation made from potato sacks. Her bosoms were two halves of an empty Spherical Dutch-cheese tin, from the galley. Two crowns cut from tin and depicting dolphins sporting on the waves were made for the 'Royal' pair and a huge set of whiskers for 'His Majesty'. The barber and his assistant made themselves minstrel outfits from canvas painted with red strips, representing a barber's pole, I made them a large wooden razor. The policemen raked up two uniforms complete with helmets from goodness knows where.

On the evening of the day before we were due to cross the Equator, Neptune (Taff Evans) came over the bows and hailed the Captain from the foc'le head and then ordered him to stop the ship. He then demanded to know the name of the ship and her business in this domain. The Captain replied that we were sailing to the Antarctic on a voyage of discovery. Neptune then inquired if there were any of his subjects sailing in the ship. He was informed there were and many others who were bursting to become subjects of 'His Majesty'. Neptune said he would hold Court on board the following day at 14:00.

All the next morning preparations for the ceremony went on. From the poop deck to the upper deck there was a drop of about three feet. On the upper deck a spare top-gallant sail was rigged to form a large tank which was kept filled with sea water by means of a hose from the fire and bilge pump. On the poop, over the tank, a tall office stool was placed from which the initiates were easily tilted into the tank, for the attention of the 'Bears'.

At the due time, Neptune and his Court formed up on the foc'le head and were received by the Skipper who escorted them to the 'Royal Throne', the engine room skylight where King Neptune and his Consort (Browning) were enthroned, guarded on either side by a policemen with drawn truncheons (Tom Crean and Lofty Hooper)

The Registrar (Tiny Abbot) in wig and gown, the Doctor (Lieutenant Bowers) in professional garb, with a carafe of medicine and bag of pills, the Barber (Alf Cheetham) and his assistant (Bill Heald) in minstrel suits and tall barbers pole hats, and the four 'Bears' (Captain Oates, Doctor Atkinson, Patsy Keohane and Joe Ford) then took up their appointed places to carry out their 'duties'. The Registrar read out the first name on the list – Doctor Nelson – Neptune instructed the police to arrest him and bring him before 'Their Majesties'

Doctor Nelson gave them a run for their money, round the deck, up the rigging, till at last he more or less allowed himself to be caught and taken before King Neptune for interrogation. 'His Majesty' asked him if he had ever crossed the Line before. Receiving a reply in the negative he instructed the policemen to take him to the stool for the initiation ceremony.

He was first examined by the Doctor who, with a two pound hammer, tapped him none too gently on the chest, at the same time telling him to repeat 99. Then with a special thermometer, about two feet long, made from a broom handle, which he jammed into the victim's gills, he took his temperature. Withdrawing the thermometer he pronounced it a very bad case, twenty below zero, and then and there prescribed two pills and a dose of medicine. Pills were about the size of West Country 'dough boys', made of dough and filled with all sorts of things of a peppery nature. The medicine was a mixture of vinegar, pepper, mustard and what not. By the time he had one pill in each cheek there was hardly room to get the medicine, which was administered by one of the wooden cooking spoons from the galley into the victim's mouth.

Then he was handed over to the barber and his assistant, each armed with a distemper brush, one with a pot of white lather which looked uncommonly like paste and the other a pot of black, the principal ingredient of which was Stockholm tar.

They set about him with a will, all the while asking him such questions as 'have you ever seen a mermaid?' or, 'what did you get in the net last night?' If he was unwise enough to open his mouth to answer, in went one of the brushes or at least as much as would go in! Suddenly he was pushed over backwards into the tank to wrestle with the 'Bears' who soused him well under water. By the time the 'Bears' had finished with him he hardly had sufficient breath, or strength left to climb out of the tank which was by way of the under side of the main rigging.

I had already crossed the Line many years before and thoroughly enjoyed watching the discomforture of the initiates. Life is, however, full of surprises and when all the initiates had been put through the mill the two policemen, grinning from ear to ear, grabbed me and hauled me before Neptune. His Majesty asked if I had ever crossed the Line before, I said I had, which was of course, true. Then he asked me if I had my certificate with me, I was sunk! I had to admit I had not. He, however, was most sympathetic and told me in a very oily tone of voice that I would get one this time and I was to be sure never to go to sea without it again. So I was put through the mill.

I gathered from the remarks of the barber and his assistant, as they slapped all that remained of the lather over my face, head and chest, my crime was that I did not serve out enough washing water and they were getting their own back. Just as the barbers were about to finish me off Lieutenant Gran who was particularly robust, rushed along the poop with arms outspread and swept all within reach, including myself and the barbers, into the tank on top of the 'Bears'. I just managed to get my head above water, under somebody's arm, and caught a fleeting glimpse of the wooden razor and Alf's tall hat floating around amid a struggling tangle of arms and legs.

Suddenly there was a crash of splintering timber as the spar holding up the far end of the sail gave way, leaving us high and dry, a heap of struggling humanity, trying to sort ourselves out. At the close of the ceremony Neptune called the Captain's attention to the chafing of the main brace which he ordered to be spliced immediately. The Captain had the 'material' for splicing on the top line. The stewards handed out a good tot all round and so ended the ceremony of crossing the Line, it was great fun. In the evening all heads, officers and men, had a singsong on the foc'le.

Crossing the Line Certificate.

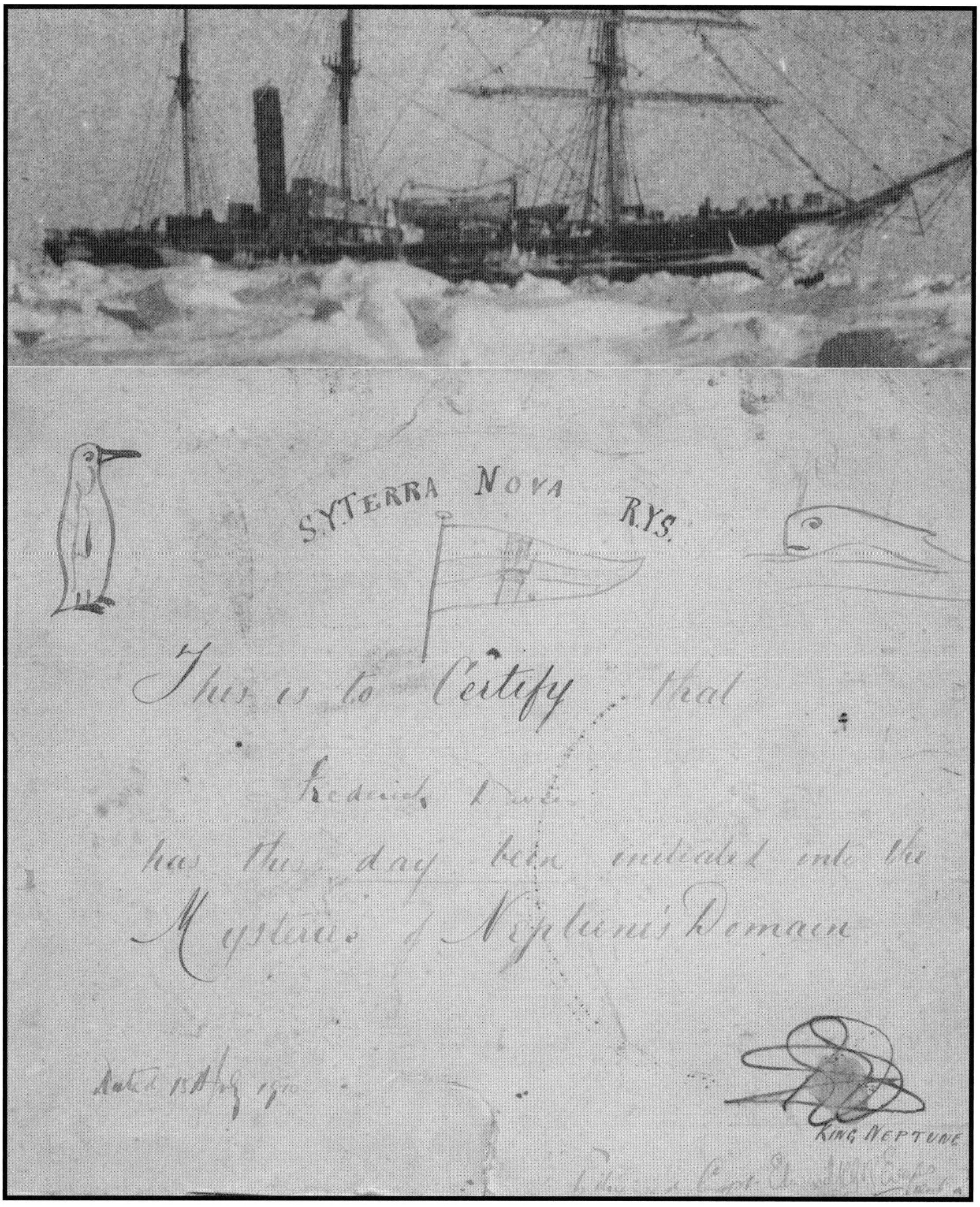

We sang all the old sea songs and shanties, I wish I could remember them all. Lieutenant Evans sang one or two of the Naval favourites. Alf his old favourite 'She Had A Dark And Roving Eye' and 'Her Hair Hung Down In Ringlets' and many others. [* note 2] Doctor Wilson sang a song of his own composition about his 'Bird, Bird, Bird, Birdies'; Bill Smythe sang a song about coiling up his ropes on shore and having a bit of a spell with his long haired pal, the girl that he adored and so on – the last thing I can imagine Bill ever doing. Tom McKenzie accompanied the singers on his mandolin.

So ended a perfect day – a wonderful day! A complete break after weeks at sea in a little ship on a great big ocean.

In order to take full advantage of the Trades and later the Westerlies, we had edged our way towards South America, and finding ourselves close to the Island of South Trinidad it was decided to drop anchor in the lee of the island to give the crew a couple of nights in their hammocks and so break the monotony of watch and watch (four hours on and four off duty, continuously) and to set up the rigging.

South Trinidad is a very small mountainous island about five hundred miles off the coast of Brazil. It was, at one time used as a penal settlement by the Brazilian Government, but at this time it was uninhabited. It got a lot of publicity at the end of the last century as a reputed treasure island, where the sea pirates of old secreted their treasure. Several expeditions had in the past been fitted out to search for the supposed treasure, though I never heard of any being found. The workings of some of the expeditions are still plainly visible.

The water round the island was as clear as crystal. Looking over the side the cable could be seen right down to the anchor. The bottom appeared to be of grey sand or pumice. A great variety of tropical fish, beautifully marked all colours of the rainbow could be seen swimming in all directions. We spent the first night with fishing lines over the side, hauling in fish as fast as we could bait the hooks. These, apart from their scientific interest were a welcome addition to our diet. We also caught two or three big sharks with the special shark hook all ships seem to carry, baited with a lump of salt pork. They were hoisted up to the cat-davit and shot. There is no love lost between sailormen and these nasty brutes.

It was intended that all the crew should have a run ashore to stretch their limbs but it was a case of man proposes. On the first day the scientist went ashore with traps etc. to collect specimens and the doctor took one of the sea men who had been sick for some time.

It came on to blow in the afternoon and it was with difficulty that anybody was able to get back to the ship through the heavy seas breaking on the rocks. The doctor and the sick man had to remain ashore until the next day, not a very pleasant situation as the island is infested with land crabs that have a pretty evil reputation and have been known to attack castaways in their thousands and clean the flesh from their bones, but on this occasion they certainly behaved themselves.

There were quite a number of Frigate or Man-O-War birds with their beautiful swallowtail tails flying around in the vicinity. Strangely enough, we had not seen a great number of birds far out to sea, possibly this may have been due to the fact that it was the breeding season. We never got another opportunity to go ashore on the island and within a couple of days we continued on our way. The seamen had been very busy during the day time and had set up all the lower rigging. They had much appreciated a couple of nights in their hammocks.

Shortly after leaving Trinidad, Mr Mate sent for me and pointed to a paint mark on the mizzen mast where it passed through the saloon pantry, it looked as though the mast had sunk about an inch. I tumbled to the cause at once. The mast was stepped on a beam over the shaft tunnel and the rigging had been well set up causing the beam to give a little.

Often during the dog watches one of the officers, or a scientist would come along to the foc'le messdeck and give a talk on their particular 'ology, birds, whales, ice Navigation or any other subject of general interest. These talks were eagerly looked forward to, and were both instructive and enjoyable and appreciated by all for'd.

Well south of the Line we saw our first Albatross. There was just an odd one or two at first the largest species, The Giant Wanderer. What magnificent birds they are, many of them with a spread of wing up to fourteen feet, snowy white, with black on the upper sides of the wings. They plane endlessly around and around the ship in quite an effortless manner, rarely moving their wings to fly, following the ship for days on end. At times they settle on the water astern, carefully folding their wings which appear double-jointed as they do so. They look like big geese as they rise and fall with the sea, feeding on the garbage from the ship. No matter how rough the weather it doesn't bother them in the slightest.

The Albatross is never seen north of the Equator. I don't suppose they would be able to live in the windless area of the doldrums, possibly this is the reason why they are never seen north of the Line. They like the windy areas, the more wind the better they like it. Their breeding places are the Lonely Islands in the Southern Ocean. From the laying of the eggs until the young are able to fend for themselves is eleven months. I don't know how often they breed but I can't think it is annually as they are met so many thousands of miles from land and must stay at sea for months on end.

Sailors always say the Albatross are old sailors, and are very superstitious about killing one of them, they feel sure disaster would overtake the ship. When they rise from the water they cannot fly right off like most sea birds but have to run along until they are lifted off by the wind, like an aeroplane, or from the crest of a wave. In spite of superstition a few were caught and preserved as specimens by the ornithologists. They were caught by the simple method of trailing a triangular piece of tin astern on the end of a line, baited with a piece of salt pork. The tin had a triangular hole cut in the centre, large enough to allow the bird's beak to pass through when it pecked at the bait. When once its beak was through the hole in the tin the strain on the line held the albatross fast and it was easily pulled on board. Once on deck it was more or less safe, for unless it could get a good run or climb through the rails and fall overboard it could not fly off. They were, however, generally chloroformed immediately and then skinned by the taxi-dermists.

On one occasion after one of the birds had been chloroformed, one of the officers happened to come along whilst there was still a slight movement of its wings. He said to the man at the wheel, 'Is that bird dead, Browning?' Brownie who was a bit of a wag said, 'No sir, it's moving its wings.' 'Oh, that may be so', said the officer. 'Haven't you heard of a fowl running around after its head has been chopped off?' 'Yes Sir, but you've never seen it peck corn after its head's off.'

Many other species of sea birds were caught by trailing wax sail twine in the wind from the mizzen rigging which became entangled in their wings as they flew around the ship.

Doctor Wilson was an exceptionally clever artist and was to be seen almost daily making sketches, or water colours of the birds or a particularly fine sunset or sunrise, or maybe a fine cloud effect.

Doctor Lillie was also an artist and did many fine caricatures of members of the crew in water colours particularly of those whom nature had 'blessed' with distinctive personal adornments. As we got further south and left the warmth of the Tropics behind, the weather became increasingly bad with heavy seas which broke on board and kept the deck continually awash. There was water, water everywhere but precious little to drink. We were all rigged in sea boots, oilskins and s'westers with the trouser legs and sleeves tied with spun yarn to keep out some of the water. The ship was getting light, for we had burnt the greater part of our cargo of coal, and rolled heavily. It was very wearying, day in and day out with all our heavy sea gear, particularly if there was a good deal of work aloft.

The decks were leaking badly with the working of the ship. Our little mess was right under the seamen's heads (WCs) and the water, we always hoped it was water trickled through the deck on to the table. I fitted a sheet lead tray on the deck of the heads, but this didn't completely cure the nuisance so we hung paint pots under the worst of the leaks. If these were not empted regularly, and not infrequently they were forgotten, somebody would get the over spillings in his plate or down the back of his neck at meal ties, much to the amusement of those who were luckier.

The weather got so bad that often both watches, which meant all hands would be required on deck to hoist up the yards or clew up the sails. The seamen generally turned in all standing, often without even removing their sea boots, in order to be ready to turn out at a moment's notice. They slept with one ear listening. Any unusual sound was enough to bring them on deck without waiting to be called, for all knew that the difference, even of a few seconds, can easily mean the difference between safety and disaster at sea in a sailing ship.

Nearly all the work pulling on ropes in a sailing ship is done to a shanty (sea song). This is not due to any exuberance of spirits, for I fear one would have to be something more than an Archangel to want to sing for joy when he has just been turned out of his hammock, half awake and perhaps greeted, as he appeared on deck by a heavy sea and carried into the scuppers.

Hoisting the topsail and top-gallant yards by hand is a very heavy job, and a shanty is almost necessary in order that all shall pull together. A good shanty man is half the work, he sings a line or two, then all hands join in the chorus and pull at the same time. One favourite shanty was 'Ranzo', it will serve to explain what I mean. The shanty man would sing the line 'Ranzo was no sailor' then the remainder would chorus 'Ranzo, boys, Ranzo', giving a pull on the rope each time they sang 'Ranzo, then wait for the shanty man to sing the next line, and so on.

Sometimes the shanty man had to make up the lines as he went and they were often spicey or topical, which reminds me of a story I once heard of an old sailor who was invited to a church do in the village hall. One of the items on the programme was sea shanties, sung by the village quartette. The Vicar turned to the old salt and said, 'I suppose these sea shanties bring back memories of your days at sea.' 'Sea shanties,' said the old shellback, 'I thought they was 'yms.'

Often we would have to hang on for dear life whilst hauling on the lee-braces as the ship rolled and buried her lee-rail in the sea, entirely swamping us for seconds on end. We had to hang on to the brace at all costs. Generally the seaman on the end of the rope prepared for this and kept the rope on the pin, ready to take a turn and secure it in a split second.

On one occasion we were caught aback by a sudden change of wind, the sails all flat against the mast, luckily, without doing any serious damage. Ships have been known to have their sticks (masts) whipped out of them when caught aback suddenly.

Section of drawings for the refit of Terra Nova from whaler to expedition ship.

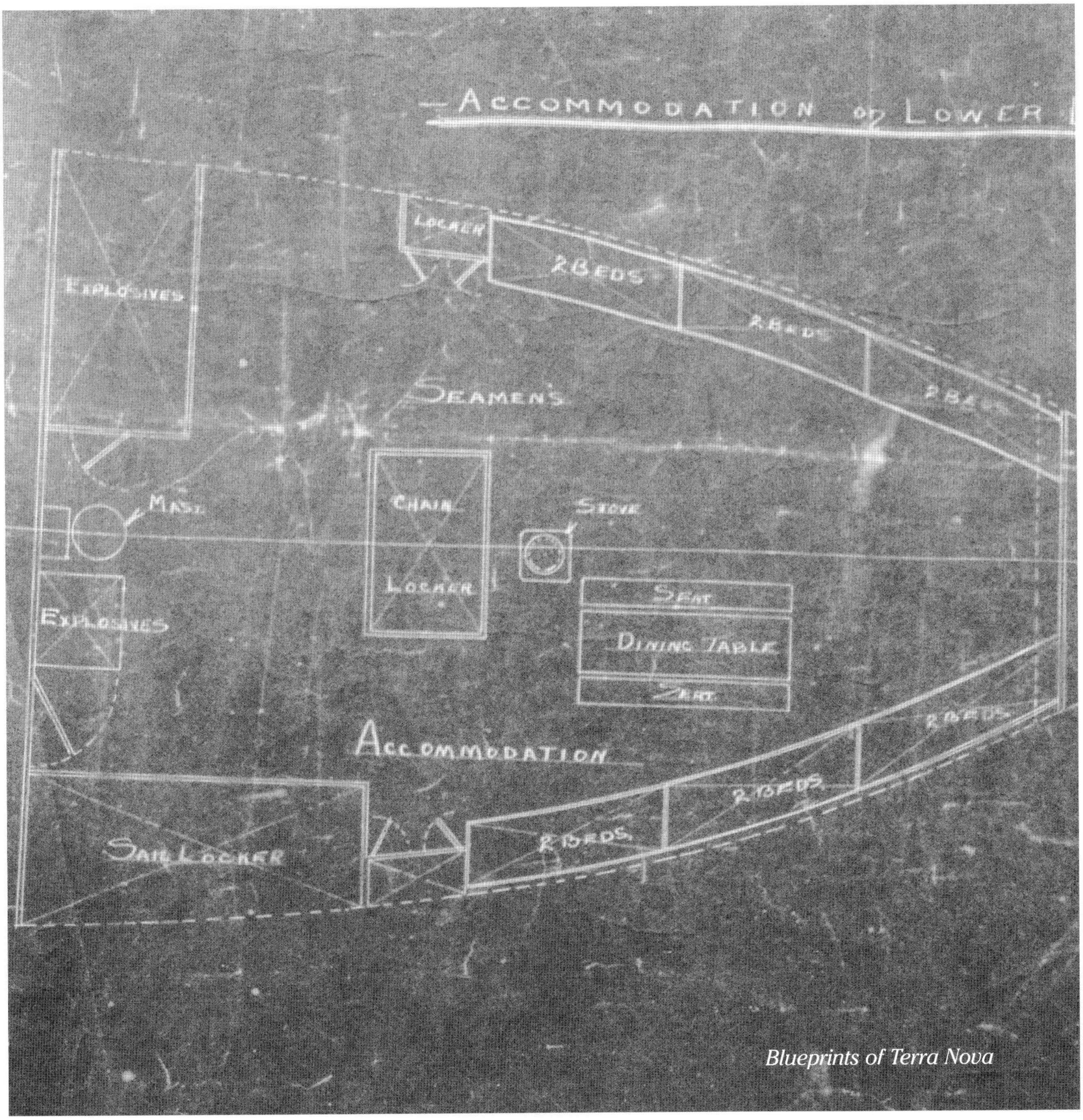

Blueprints of Terra Nova

The late **Captain Scott's Tribute to OXO**

Winter Quarters, Cape Evans, *23rd Jan., 1911.*

"I have pleasure in informing you that the various preparations of Lemco "and Oxo have been found very useful and suitable for our camp purposes.

"I enclose some photographs showing your goods in use in the Antarctic."

Oxo also formed part of Sir Ernest Shackleton's South Polar equipment and proved its worth so handsomely that Sir Ernest cabled from New Zealand:—

"Found Oxo excellent on sledge journeys and throughout winter."

Oxo is chosen because Oxo is real nourishment—better nourishment than ordinary meat, because none of the strength Oxo creates is lost in the process of assimilation.

Oxo is not merely for cases of emergency and difficulty—it is an exceedingly valuable addition to the ordinary diet.

All who value stamina and fitness for every day should follow the example of those who have made diet a study and drink Oxo regularly.

OXO, Thames House, London, E.C.

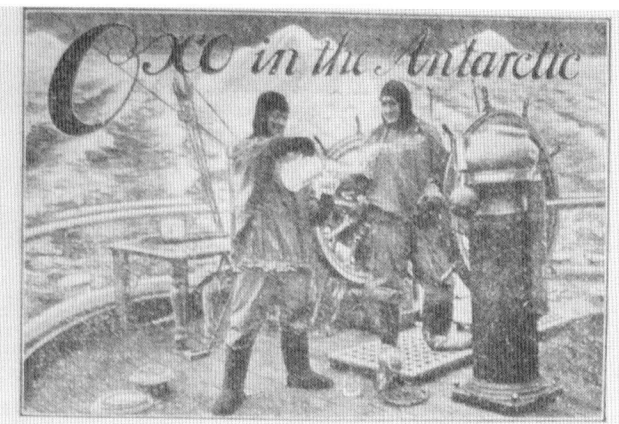

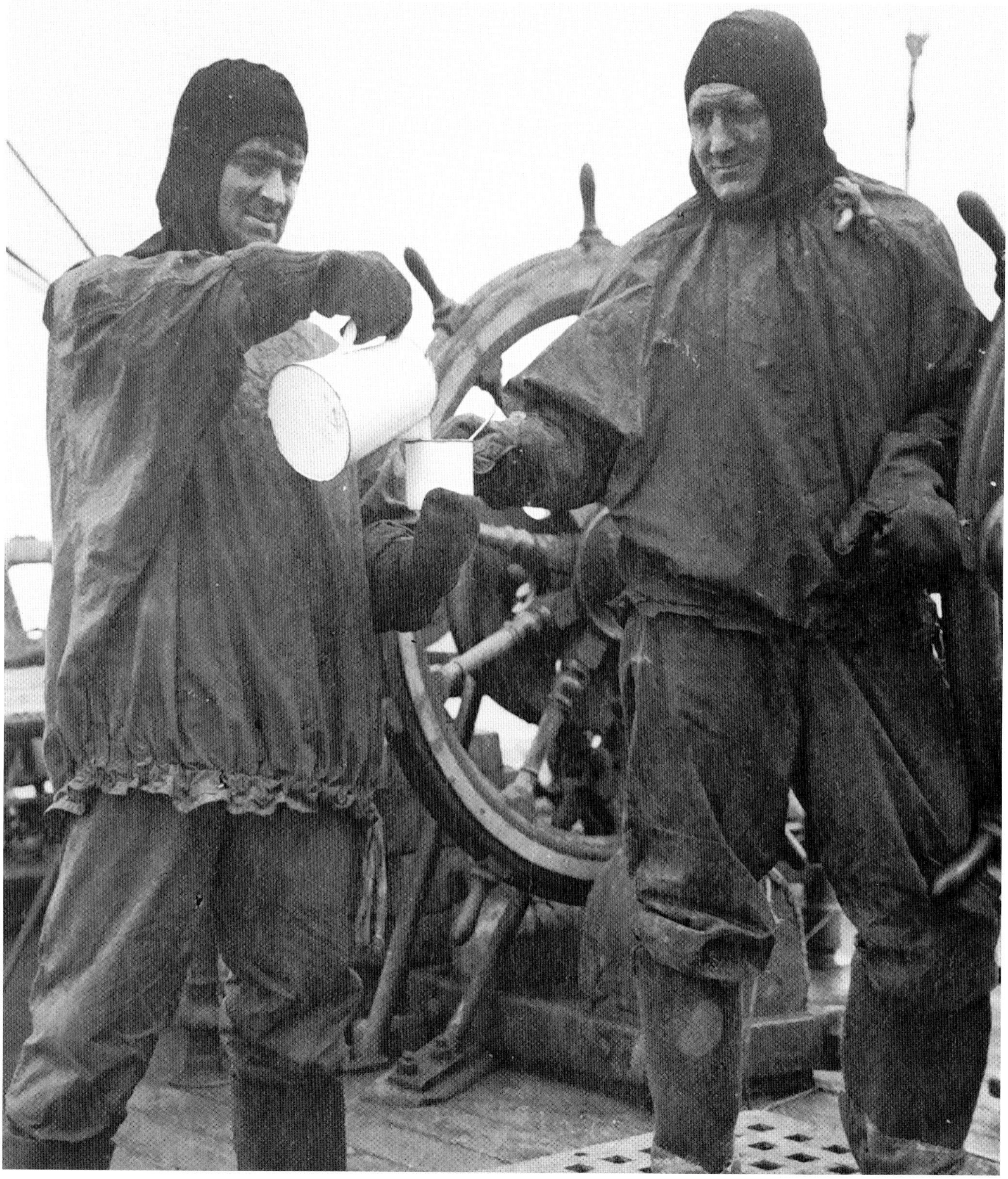

Chapter VI. Hospitality at Simonstown

Whilst we were in the south easterlies, Bill Smythe 'sails' came to me one day and confided he had found a couple of cases of stout in the 'tween decks', which nobody seemed to know anything about, and that they had already been broached. Probably they had been stowed there on the quiet by the stevedores who loaded the ship in London, but who apparently had not had time to get away with the lot. With a knowing look on his 'clock', Bill suggested we ought to do something about it, he didn't want to see anybody get into trouble, as he put it.

Barkus being by no means unwilling to do a good turn in this particular line, we discussed ways and means. We decided I should shift my bench from the upper deck to the 'tween decks, where Bill would fit it with a canvas screen so that under the bench could be used as a cellar. The bad weather offered sufficient excuse for this change of residence and was no sooner said than done. The stout was in a very lively condition and we had to give it special treatment.

For this purpose we 'pinched' one of Cookie's enamel bowls which we kept under the bench. When all was quiet in the forenoons we would open about four bottles and empty them into the bowl to liquefy. As soon as the stout was ready we emptied it into two pint mugs and put a couple more 'soaking'.

As Bill stitched away on his sails – home-ward bounders, three stitches to the inch he spun me yarns of his experiences while sailing under the 'Red Duster' (Red Ensign) after leaving the Navy, telling me tales of the time he tramped through New Zealand, doing odd jobs on farms, cutting grass, harvesting, sawing wood or any old job. On one of the farms the farmer asked Bill if he could plough, he of course said he could, so the farmer set him to work. Some hours later he came along to see how Bill was doing. When he saw the fist Bill was making of it, he told him to go back to the house and get a book, then find a nice shady tree and do a bit of reading, he, the farmer, would be saving money.

Bill was a rolling stone and had certainly gathered no moss. During the dark hours he dumped the 'dead marines' (empty bottles) over the side, and later the cases, nobody got into trouble thanks I suppose to our 'kindness' and nobody was ever the wiser.

My bench soon became the meeting place where a select few came to spin their 'benders'. Alf, the Bos'n, brought his paint mixing outfit here. My mates Captain Oates, and Doctor Atkinson took great delight in Alf's yarns and would encourage him to sing some of his 'special' shanties, one in particular described what Alf saw through the key hole in the door.

He told his tales as he mixed the paint, when he had mixed it to his own satisfaction he would ask me what I thought of it. I would examine it carefully in a most professional manner, and then suggest he ought to add a little more turps or linseed oil.

After he had acted on my advice he would again ask me what I thought of it. I would tell him that he had overdone it, and that it would now need some white lead to thicken it. Finally Alf would apply his own infallible test, if the stirring stick remained upright in the mixture and did not fall into the side of the drum it was as he said, 'a good drop of stuff'. At times I would hear one of the after guard calling out to Alf for a pot of paint with a kink in it for the davits, or a pot of spots for some particular job.

Lieutenant Pennell, Captain Oates and Doctor Atkinson were great pals. Lieutenant Pennell asked Doctor Atkinson to make a couple of rigols to fit over the port-holes of the chart-room, to prevent the rain running down the side of the chart-house into the port-holes when they were opened. Doctor Atkinson sought my advice on the matter. He wanted to know how to set about making these 'eyebrows' as he called them. I cut him a couple of strips of sheet lead, gave him a mallet and explained to him how to proceed. I told him if he beat the lead on one edge it would gradually turn to the shape of the port-hole.

For some time I could hear him banging away on the deck overhead with the mallet. Suddenly the banging ceased. Next thing I saw was Doctor Atkinson in the doorway of the 'tween decks, shaking the mallet at me with one hand and holding a lead strip, that looked more like a bootlace over one of the fingers of his other hand which he held towards me saying, 'wait until you go sick.' I told him he had not followed my instructions. He swore he had and said I had pulled his leg. I showed him again exactly how it should be done and he had another go. This time he was more successful and in due course the rigols were ready to fix to the chart-house. After he had finished the job he asked me to come along and have a look at it which I did, I could see he was very pleased with himself. 'What do you think of the job, Chippy?!' He asked. I replied, 'Well, doctor, if I hadn't known you did the job I would have thought it was an effort of prehistoric man.' And then made myself scarce in a hurry. When the ship paid off nearly three years later, Lieutenant Pennell had the rigols taken off and took them home as souvenirs.

The doctor got his own back on me pretty soon, for a day or two later I ran a big splinter under my thumbnail. I went along to him to get it removed. He asked me to sit on a stool in one of the laboratories and went off saying he would only be a minute or two. When he returned he had Captain Oates with him. He said to Captain Oates, 'I've caught the bird at last, just watch me perform.'

Before leaving London, the Chief Engineer had fitted a small evaporater for condensing fresh water. He was very proud of this, near useless bit of machinery. Every morning when we were under steam I reported to him the amount of fresh water made during the previous twenty four hours, I could have made more and better water myself – we had only made half a ton by the time we had reached South Africa, which I kept in the separate tank. Fortunately we did not require to use from this tank until the day before we reached Simonstown. When we tasted it oh horrors! It was just as salt as sea water. We had the flag flying for the water boat when we arrived, everybody was very thirsty, for tea and coffee made with salt water is not very palatable.

It was a glorious day when we dropped anchor in Simonstown Bay. Just heavenly to have the ship steady under us again after rolling for weeks on end, better still a restful night with our clothes off, without fear of being called on deck any minute. We had no news of the outside world since leaving Madeira and no mail from England since leaving Cardiff, so we were all eagerly looking forward to letters from home.

We had almost forgotten the morning paper habit, as a matter of fact I don't think we ever got a connected picture of world events during the whole of the expedition. Some strong minded people rationed themselves in the matter of home papers, opening a certain number each day in proper sequence. When we did receive our mail they usually covered fairly long periods and naturally we looked at the latest letters first to see that all was well.

I remember one particular item of news that interested me. China had declared itself a Republic. I wondered what the new China would be like, I knew the old pretty well for I had served on the China Station from 1904-6, during the Russian-Japanese war, and was smitten by the glamour of the East. We had a unique opportunity of seeing these countries, China, Japan and Korea. After the Russians were defeated, the British squadron was entertained at all the principal ports of Japan and at Dalny, close to Port Arthur, in what's now known as Manchukuo. I was in Tokyo when Admiral Togo, who had commanded the victorious Japanese fleet made his triumphal return to Japan amid great celebration.

We received a lot of help whilst at Simonstown from the Royal Navy and the dockyard. I had several shipwrights from the squadron to help me to make good some of the more serious defects, one of which was new cross trees on the main top mast. A lot of my time was taken up scrounging useful stores and material from the dockyard. One day, in the dockyard, on my way to catch a boat back to the ship, I heard someone some distance off shouting 'Chippy'. On turning round I saw Captain Oates running to catch me before the boat shoved off. He asked me if I would like to dine with him and Doctor Atkinson ashore. I said I would be delighted.

'Well,' he said, 'get hold of old Mick Crean, Bill Smythe and Patsy Keohane and make it next Saturday at Wynberg.' We had a good time.

Captain Oates knew the Cape well. He had fought in the Boer War as a subaltern, I was told he was known in his regiment as 'No Surrender Oates', and the story goes that he and his platoon, whilst on patrol, were cut off and surrounded by the Boers, who called on them to surrender. Oates reply was characteristic of the man, 'I came here to fight not to surrender,' and not till all the little party had been either killed or wounded did the Boers get them. The Boers were so impressed by their bravery that they did not take any of them prisoners.

Whilst in Simonstown, I got permission to go on board an old hulk which had been HMS *Penelope*, used during the Boer war for housing prisoners of war, to select a couple of her downton pump handles for use on our hand pump. These I got the dockyard blacksmith to alter so that they fitted right across the ship, from pin rail to pin rail. This was a very great improvement on the very short handles, fitted originally, which could only be manned by about eight hands, now the pump could be manned by all hands if necessary. I hardly dare to think what would have been the fate of the ship later on without them. An attempt was also made to trace the leak that had been giving so much trouble, which we still thought was around the stern, but without success.

The people of Simonstown were very hospitable and entertained us in their homes and clubs, while the Naval Canteen ashore became a sort of home from home, with beer a tikky (3d) a bottle and Cape brandy (Cape smoke) almost for the asking.

The Navy gave us a great send off. The night before we sailed they held a smoking concert in the canteen to which we were all invited. I well remember Lieutenant Evans' opening remark when he rose to thank them for their hospitality on behalf of the ship's company. He said he felt like the little boy caught with his trousers down, he didn't know what to say. Anyhow a good time was had by all.

During our stay the ship was for ever full of visitors, friends we had met on shore. One afternoon I had two of my friends from the dockyard on board, the cashier and his wife. Whilst showing them over the ship the cashier met with a slight accident which gave him a nasty shock. We had arrived at the saloon which was reached by a ladder from the poop. At the bottom of the ladder in a dark corner was a small hatch leading to the lazarette. This was usually kept closed but when open a brass bar was hooked across to prevent anyone stepping in. Coming from the bright sunlight it was difficult to see anything at the foot of the ladder. I led the way and stepped over the hatch without noticing it was open, the lady also passed safely over but the cashier dropped through into the inky void below. He just managed to grip the edge of the deck around the hatch, to which he clung desperately.

His feet were actually only a few inches from the floor of the lazarette, but he didn't know that of course. We got him up, he was badly shaken, but soon recovered after a tot of brandy in the saloon.

All good things come to an end and once more we were on our way with many presents from our Cape friends of home made cakes, jam, including some of the famous Cape gooseberry jam.

Portrait photograph of Captain Lawrence 'Titus' Oates, sent as a a gift to 'Chippy'

Chapter VII. The Roaring Forties

We had a pretty tough time in the Roaring Forties, gale succeeding gale almost without a let up. There was so much pully haully that the Shanty men soon exhausted their repertory and were repeated over and over again 'Ranzo', 'Way Down Rio', 'Shangi Brown', 'My Old Man's A Fireman In The Elder Dempster Line', and the old *Terra Nova* boating song, 'Pumping, Pumping, Pumping, Always Bally Well Pumping,' to the tune of a well known hymn.

The hand pump was giving more trouble with the continuous bad weather. I spent hours in the bilges cleaning the suctions, the bilges were reached through the after watch and then down through a very narrow wooden trunk with rungs on the after end. When the seas were breaking badly on deck it was difficult to get the after hatch open and we had to wait an opportunity when there might be a lull, to uncover the hatch so that I could be popped into the trunk, bucket on one arm and one electric magazine lamp on the other, cluttered up with oilskins and sea boots. The lamp was a rather heavy affair carried by a big strap which allowed the whole thing to be inverted to switch on the light, at least that was the idea for more often than not it conked out, it was not constitutionally strong enough to stand up to such a hard life.

Once inside the trunk I made my way to the bottom, some twenty five feet down until I stood on the keelson all in total darkness. I then placed my bucket in one corner and switched on the magazine lamp which would, if I were lucky show just a glimmer of light. After lashing the lamp to one of the rungs I had to wriggle myself into a position with my legs up the trunk and my shoulder on the keelson, I could then reach the bottom of the pump suction between the frames.

If there was not a great deal of water in the bilges I managed to get hold of the coal-balls that had accumulated around the suctions quite easily and put them in the bucket. When the bucket was full I had to take it to the top of the trunk to get it emptied. On the other hand, however, if the water was over the keelson as was more often the case, the bilge water swished over my head and face covering me in grease and coal dust.

When I had cleaned the suctions I climbed to the top of the trunk with my bucket and lamp and banged on the hatch, which would be opened by the waiting seaman at the first opportunity, when I would emerge looking like a Kentucky minstrel. I removed the muck in the lamp room with a piece of waste soaked in paraffin, there was no fresh water for washing purposes.

This process was repeated many times during the day and night, more frequently in very bad weather. Possibly by the time I cleaned the pump suctions it would be necessary to strip the pump again to clean the valves. I can hear some folks saying, why didn't the B.F. take a torch or a wandering lead with him down the trunk. The simple reason was these items were not so common in those days, beside we had no dynamo and so no electric light. The ship was lit between decks I said lit, by paraffin lamps and sometimes candles, which resulted in semi darkness. No lights are ever allowed on deck in a sailing ship, not even through the chinks of the skylight, as the officer of the watch would not be able to see what was happening aloft.

In howling gales with thunderous seas breaking aboard, the seamen worked at night, on deck or aloft, in inky blackness, except perhaps for the light of the moon. That's why everything in a sailing ship must be all ship shape and Bristol fashion. Each rope has its particular pin and the halyards coiled down on deck ready for letting go at a moment's notice, at the same time so secured that they cannot be washed about by the sea and become foul. The old salts do everything by touch or is it instinct?

My job, whilst the seamen were aloft, was to attend the ropes on deck. This had to be done almost by instinct as it was impossible to hear any orders however loudly they were shouted, even by anyone standing quite close let alone from aloft, when the gales howled through the rigging. This was not made easier when the canvas and ropes were half frozen and the ship rolling to an unbelievable angle. It was really frightening, how much we appreciated the coming of the dawn after such a night.

We were hoping to land on the lonely uninhabited Island of St. Paul, to get a goat or two for fresh meat. Captain Oates and I were busy repairing the Norwegian Pram, a very handy boat for landing through the surf. It had been smashed on deck by very heavy seas. It did not materialise, however, as the day before we were due to arrive we had a particularly heavy gale and did not as much as glimpse the island as we scudded before it. In spite of bad weather we were always cheerful, there being many hands to work the ship and the difficulties with the pump were now taken more or less as part of the normal routine.

There were hundreds of birds around the ship including several species of Albatross from the Giant Wanderer to the beautiful Sooty which is considerably smaller, Cape pigeons (petrels) and the tiny stormy petrel that never seem to get into the water but just to touch it with their feet as they fed on the small crustacea.

Billy Neald, the steward, caused some amusement one day, when he came for'd just after Lieutenant Pennell, the navigator, had fixed the noon position saying he had just seen the position on the chart and we had only about that much to go, as he measured the distance down his forefinger with his thumb, what matter that this represented nearly another two thousand miles!

One of the firemen had a tame squirrel that had been given to him by a keeper at Groote Schuur, near Cape Town, the home of the late Cecil Rhodes. He kept it on the mess deck in a cage he made from a large packing case. During a conversation he told me that he could not understand why his squirrel was so wild and frightened whenever he went near it, as it was quite tame when he got it, some of his messmates could have told him why! He asked me if I would have a look at it as it had a sore on the end of it's tail. Sensing a bit of fun, I said I would.

So off we went to have a look at it, it was to say the least a little more than shy. I asked him if he had had it vaccinated before bringing it aboard. He replied, 'No, they don't vaccinate animals, do they?' 'Of course they do.' I said. 'Weren't you vaccinated when you joined the expedition?' He agreed he had been. 'Well,' I said, 'It's the same with animals.' He looked very down in the mouth, but brightened up when I suggested I would mention the matter to Doctor Atkinson if he would care for me to do so. He said he would be very grateful if I would, and thanked me. I told Doctor Atkinson what was in the wind and he said he would come along to the foc'le messdeck at 16:30 to have a look at the squirrel. I passed the word to both watches.

Somebody printed a large notice and stuck it up just outside the foc'le door. It informed all and sundry that 'Professor Billy', the famous lion tamer would enter the lions den at 16:30 and invited them to come in their thousands. A big audience had gathered by the time Doctor Atkinson came along. Billy opened the cage door very cautiously and with his bare hand made a grab for the squirrel. At the same time the squirrel made a grab for Billy's hand and won by a neck, sinking its teeth into one of his fingers, amid great cheering from both watches as he tried to shake the savage animal off. Somebody produced a leather glove and he had another go, this time with more success.

The doctor had a look at the sore place on its tail and expressed the opinion that it was caused by jumping about in its cage, adding that he didn't think it would be necessary to vaccinate just then, but would let him have some ointment for it. Billy came to me afterwards and solemnly thanked me for bringing the doctor along. I told him not to mention it, it had been a real pleasure to perform so small a service. On arrival in Melbourne I believe he presented the squirrel to the Zoological Gardens.

A land fall is a great event after weeks at sea. All hands are on the look out, each hoping to be first to see it, and excitedly anticipating a few days ashore. Paint brushes and scrubbing brushes work overtime to make the ship look as spick and span as possible. Somehow in the last few days mast and yards get painted, rigging blackened down, deckhouses all shining white in new paint and even the decks have been Stockholm tarred and sanded.

We made a grand picture bowling along under full sail and steam in calm sea, as we passed through Port Philip Heads making for Melbourne. At last we dropped anchor at Port Melbourne. As soon as the anchor was down I had a job aloft to repair the lightning conductor on the top-gallant mast. Captain Oates and Tom Crean gave me a hand. From the mast we had a fine view of the harbour. Pleasure steamers and motor boats were all around the ship making holiday. How lovely it was in the warm sunshine with no wind and the harbour like a mirror. One of the motor boats, full of 'fillies', came alongside to ask permission to come on board, but we had not been given pratique by the customs, so they lay off and waited. Captain Oates asked Mick which one he fancied. Mick's fancy was the plumpest of the bunch.

A little later Captain Oates spotted Doctor Atkinson on deck, ready to catch the first boat ashore that was just about to leave the ship. Captain Oates shouted 'Jane' (the name by which Doctor Atkinson was known to the Afterguard). Doctor Atkinson gazed aloft. 'Got any money?' shouted Captain Oates. The doctor nodded, indicating the affirmative. Captain Oates was on deck in a brace of shakes, and in less than a minute later he climbed into the boat with his horse going togs tucked under his arm, and a pair of shoes dangling from his hand. That was the last I saw of my mates until we sailed three days later, I guess Melbourne Races and Captain Oates love of horses, allied to the fact that it was Melbourne Cup Day, the Australian Derby, the following day had a lot to do with it.

Whilst at Melbourne we were visited by the Prime Minister of Australia and some of his Ministers.

Shortly after sailing from Melbourne we passed HMS *Powerful*. The wind being fair, we were making plain sail as we passed her. Her ship's company gave us three cheers. We manned the rigging and returned the compliment.

Several years later, in 1942, a special constable on the dock gates at Boston, Lincolnshire where I was then serving as Boom Defence officer in command of the War Defences stopped me and said: 'You were with Captain Scott, weren't you, Sir?' I nodded, and he then produced two photographs of *Terra Nova* under full sail leaving Melbourne. He told me he had taken the photographs himself from *Powerful*, in which he was serving as a Petty Officer over thirty years earlier.

The People's Tribute

Gramophone music played on Antarctica

Mrs Scott meeting one of the ponies.

Chapter VIII. Disaster Narrowly Averted

The voyage to Lyttleton was somewhat uneventful. The weather being fine with little wind, we steamed most of the way. One incident only stands out in my memory. Captain Oates and I were over the side on a stage, repairing the bulwarks. He bet me a pint of beer, in Lyttleton, the job would not be completed that day. I told him it was money for 'old rope', and so it was.

On arrival at Lyttleton, New Zealand, no time was lost in preparing for the southern voyage. Everything was taken out of the ship and placed in a warehouse for resorting. The ship was dry docked with the hope of finding the leak.

As I was not entirely satisfied with the huts for Winter Quarters on leaving London, I transferred them lock, stock and barrel to a recreation ground ashore. Here the framework was erected and all boards cut out to proper lengths, tied up in bundles and stencilled to facilitate the erection in Antarctica. This also saved a lot of valuable stowage space in the hold for there was not a square inch to spare. I had four seamen to assist me, Joe Ford, Patsy Keohane, Tiny Abbot and Brownie. It was really a stroke of luck that the huts had to be put together again. It gave me a chance to see everything was quite in order at the eleventh hour and at the same time I was able to give the seamen a certain amount of training, for it was quite on the cards they might have to do the job in the Antarctic themselves in the event of the ship having to leave before the huts were completed. There was a constant stream of visitors to see the huts. Many of them, thoughtfully, brought bottles of beer which were much appreciated, the weather being very hot at the time.

Beside the stores we had brought out in the ship, a considerable amount had arrived by a cargo vessel. All these were collected together in the warehouse close to the jetty, where the ship was loaded, sorted and checked, so that they were in proper order for unloading in the Antarctic.

Twenty ponies, all grey and sixty sledge dogs had already arrived and were quartered on Quail Island, the leper island in Lyttleton harbour. These had been brought from Siberia by Mr Meares and two Russian boys, Anton for the ponies and Demetri for the dogs. Captain Oates took charge of the ponies on arrival. The seamen had to turn out of the foc'le mess deck in order that the stalls for fifteen ponies could be fitted there. Another four ponies were accommodated in stables built on the portside of the upperdeck, between the galley and the icehouse. Gradually everything was stowed aboard. All the main hold was taken up by four tons of coal and thirty tons of briquettes, the forehold was packed with cases containing food, in the tween decks there was a smaller locker of explosives in case we had to blast our way through the ice, all the timbers for the huts and remainder of the cases and compressed fodder for the ponies. Under the 'tweendecks were four water tanks, three of which were filled with oats for the ponies, provided quietly by Captain Oates at his own expense. This left only one tank of drinking water about twelve and half tons, which had to last until we reached the pack-ice before we would be able to replenish by melting ice.

On the upper deck, in addition to the ponies there were three motor sledges in wooden cases lashed by chains to the deck, one on either side of the main hatch and one on the after end of the hatch. These formed a kind of zareba that offered some protection for a few of the dogs. On top of the after sledge, and against the main mast, bales of hay were stowed reaching many feet up the mast.

Captain Oates with some of his charges on board ship

The remainder of the space on the upper deck was taken up by cases of petrol and carbide which were securely lashed.

In the ice-house, two hundred carcases of mutton and five of bullocks were stowed on a couple of tons of ice. Many more carcases were lashed to the mizzen rigging. Dog and man hauled sledges, ski and ski sticks were lashed on top of the saloon. The tiny cabins were piled high with the scientific instruments and personal belongings of the three, and even four occupants. All the spare spars, spare rudder and spare propeller were landed to give more space for stowage.

The day before we sailed the ponies were got on board. They had to be humoured one by one into the foc'le stalls, and there they were forced to remain until landed. As each pony was put into its stall it shut in all the ponies put in before it. The hawse pipes were cemented up and bales of hay packed right forward around the heel of the jibboom, as far back as the first stall. It was an easy matter to get the last four ponies in the stables on deck.

Then came the dogs. Some were chained in the shelter of the motor sledges and the remainder anywhere possible on deck, fore and aft the ship, one was even chained to the windless, which had almost disappeared under an assortment of odds and ends, on top of all this was a small hutch containing a wild rabbit that had been given to Mick Crean.

By the time the ponies had been got on board, fed and watered and made comfortable, it was pretty late in the evening. Captain Oates then said to me, 'What about that pint of beer I owe you, Chippy?' I thought it a good idea and said, 'I consider we have both earned one.'

We went along to a hotel, just a few minutes from the ship, Captain Oates in a grey cardigan and muffler and his famous canvas trousers, which he had made himself under the instructions of 'Sails', and of which he was very proud. We had a couple of quick ones and returned to the ship. He would not leave his charges for long. He told me how relieved he was to get the ponies safely stowed away on board, and said he had been so anxious on the previous night that he had been unable to sleep and filled in time soleing a pair of boots.

The shipwrights had made a good job of the stalls and stables, but I regret to say they were unable to find the leak.

We sailed on Saturday afternoon, 27 November. This gave the public an opportunity of seeing the ship off. They came by train in thousands, the jetty was packed to capacity. Amongst them were many friends of the ship's company for we had been treated with great hospitality by the New Zealanders. Just after midday, a short service was conducted on the poop by the Bishop of Christchurch. It opened with the sailors' hymn, 'Eternal Father Strong to Save', and a prayer for the preservation of the expedition from the dangers of the sea and the violence of the elements, followed by another hymn, 'Oh, God Our Help in Ages Past', and ended with a short address by the Bishop.

After the service the pipe band of the Christchurch Caledonian Society played on the poop. We had many pleasant evenings with the members of the Caledonian Society, both at their headquarters in Christchurch, and on the ship.

The ship was crowded with visitors almost to the last moment, all most interested in the ponies and dogs. At 14.30. it was 'All Visitors Ashore' and in a few minutes the hawsers were cast off and a tug pulled our bows away from the wharf towards the entrance of the moles. We then proceeded under our own steam.

There was great cheering from the crowds on the jetty and ships in harbour blew their sirens, Craft of all description accompanied the ship as far as the Heads. On one of the tugs was the Garrison band and on a pleasure steamer our old friends the pipers. There was a special cheer for Doctor Atkinson from the stevedores with whom he was very popular, having worked with them loading the ship. At the entrance to the Head, HMS *Cambrian* was at anchor and cheered as we passed.

An amusing incident happened as we cast off from the jetty. Bob Ford was the helmsman, and one of the scientists had been detailed as lee-helmsman. The helmsman received an order from the bridge and turned the wheel. When the scientists felt the wheel move he clung on like grim death, with the result that he was pulled over the wheel. The spectators on the jetty thought it was all part of the performance. Once clear of the Heads we shaped course for Port Chalmers to pick up a gift of coal.

Captain Scott and his wife returned to Lyttleton in the tug and travelled to Port Chalmers by train. All hands gave three cheers for Mrs Scott as the tug left us.

During the time we were at Lyttleton there was an agricultural show at Christchurch, and as there would be no work on the ship that day we were given the day off.

The previous evening at one of the hotels, a Dutchman got into conversation with me. He told me he was a wrestler and had a booth at the show and wanted to know if there was anyone on board who could wrestle. He was looking for someone who would be willing to give a few exhibition bouts with him. I thought of Tiny Abbott, who was a physical training instructor in the Navy and a good all round athlete, and mentioned the fact. He persuaded me to try and arrange the matter with Tiny and bring him along to the booth.

On board that evening I broached the subject to Tiny, we thought it would be the making of a good day out, so it was arranged that a small party consisting of Tiny, Patsy Keohane, Brownie and myself should visit the show. The following morning we were on our way, bright and early for Christchurch, it was a lovely day to 'bust' a quid'. At the show ground we looked around for 'Dutchie's' outfit. We spotted him outside his booth rigged in the usual trimmings of a wrestler, weighed down with medals, offering to take on all comers.

I caught his eye and he motioned us to the back of the stage where I introduced the party. He arranged with Tiny to give a few exhibition bouts to draw the crowd, but kept emphasizing the fact that he didn't want Tiny to throw him as it would spoil his show. Tiny, suitably attired, was taken by 'Dutchie' on to the stage in front of the booth and introduced to the crowd. 'Dutchie' was a great showman and shouted in a very loud voice 'Roll up! Roll up! We got the T'outh Pole a T'outh Pole', he couldn't say south 'Come and see him, come and see him wrestle with the Great Hackensmidt Roll up! Roll up! Roll up!'

He repeated this several times and the crowd did roll up, in their hundreds. We were now part of the show, more or less acting as Tiny's seconds. The first bout, the best of three throws ended in a draw! Between bouts we adjourned with 'Dutchie' and a few of his hangers-on to the refreshment tent where we were regaled with the free beer. After a few quick ones we returned to the booth where the performance was repeated. The result was always the same – a draw – and after about four bouts we pushed off to see the rest of the show. The affair caused a lot of fun for many a long day. In the height of the gale or the most awkward situation somebody would shout 'Roll up! Roll up! Roll up! We got a T'outh Pole a T'outh Pole, come and see him.'

Gerof Demetri Dog Handler

We arrived at Port Chalmers the next evening, Sunday. The docks were crowded with people as we made fast. We were to take another hundred tons of coal in bags, on top of the cases of petrol and carbide. As a matter of fact we could only take about eighty tons and this brought the cargo on deck level with the rail. Just before we sailed on the Monday afternoon, I topped up the one remaining fresh water tank and the animals were given as much as they would drink, everything that would hold water was filled.

Lieutenant Bruce RNR, Captain Scott's brother-in-law, Mr Meares, in charge of the dogs, the remainder of the scientific staff, Anton and Demetri the two Russian boys and several seamen, to replace those who were to be landed with the shore parties, joined the ship in New Zealand.

We sailed about 16:00 on the afternoon of 29 November 1910 and again had a great send off by the people at Port Chalmers.

For the first day or two the weather was in our favour but by the third day it deteriorated and soon reached gale force. Working the ship in the dark was very difficult. We had to man the ropes crawling over the bags of coal with practically no protection from the heavy seas.

About the third night out, whilst sweating up the fore braces, I was washed along on top of the coal bags and had the skin torn off the knuckles of my right hand. What with mucking about in the bilge water and the filthy conditions generally, my knuckles began to fester. I showed them to Doctor Atkinson, he advised me to get them thoroughly cleaned, which I did using a deck scrubber and a little hot water from the galley, rather a painful process. The doctor then applied new skin and I had no further trouble.

As the wind increased, sail had to be reduced, until eventually it reached gale force and we had practically to heave to under steam only. Then things began to happen on deck, for in spite of the care with which everything had been secured, some of the deck cargo was torn from its lashings by the heavy seas. I was busy on the Port side amidships, securing some of the loosened cases abreast of one of the motor sledges, when a huge sea struck the ship. I hung on for dear life and remember looking up through the water, it was green and clear. When it subsided I found I was trapped by the foot and about twenty feet of the bulwarks had been washed away. I had to wait until the next heavy sea lifted the timber off my foot. This same sea washed two of the dogs overboard, clean through their collars, one of them was washed back aboard again and saved.

Whilst I was trapped by the foot, one of the leader dogs, 'Vida,' a particularly savage beast chained close by, caught me by the sleeve and tore it off. One of the seamen always declared that he saw me washed overboard with the bulwark and washed back again, hanging on to the fore brace.

Since leaving Port Chalmers we had been under sail and steam in order to reach the shelter of the pack-ice as soon as possible, on account of the animals who were having a very bad time.

When under steam the bilges were pumped out by the steam pump in the engine room and it was not necessary to use the hand pump. On the second night of the gale the bilge pump in the engine room choked. The engineer on watch, instead of reporting to the bridge that this had happened hoped I suppose, he would be able to clear it. However, he did not get it cleared and by the morning stokehold plates were washing about the stokehold.

By this time the gale had reached such force, and there was so much water on deck from the tremendously heavy seas breaking on board, that it was impossible to open the after hatch to get at the hand pump suctions and little more than a dribble of water was coming from the pump. By eight o'clock in the morning things were really serious, the water was under the boiler and fires had to be drawn.

The ship was rolling drunkenly, when one of the doctors and I fetched a small downton pump from for'd, to use in the engine room. We had struggled along the deck as far as the break of the poop when a big sea came over the rails and we floated away, pump and all, rolling from side to side of the ship, until we were grabbed by some of the hands manning the hand pump. When we eventually got the pump to the stokehold it was of little use.

Captain Scott and Lieutenant Evans discussed the situation and came to the conclusion the only thing to be done was to cut a hole through the steel bulkhead between the boiler room and the main hold, in order to get at the hand pump suctions, and in the meantime to bale out with buckets. Fortunately I had just the tool for the job, a driver for cutting a steel plate. Bill Lashley and myself commenced to cut the hole immediately, lying on the hot boiler on wet sacks to keep the heat off our bodies. Lieutenant Rennick offered to help but I was afraid he might drop the tool behind the boiler, then we should have been done.

While we were cutting the hole, a baling party of officers and scientists formed a chain up the stokehold ladder, passing the buckets of water from hand to hand from the stokehold to the poop. At the same time the hand pump was kept going, in spite of the small amount of good it was doing, but every little counted, for it seemed the ship might founder at any moment. All this was not so easy as it might sound with the ship rolling like she was, but there can be no question whatever that this alone kept her afloat, until the hand pump was got going again.

It took about ten hours to cut the hole in the bulkhead, we then found there was a three inch wood lining to cut through. After this it was a comparatively simple matter to dig through the coal in the hold, to the pump well trunk, from which I removed a few boards so that we could get into the trunk. Lieutenant Evans went in first and I followed. There was so much water that it was difficult to reach the bottoms of the suctions we practically had to dive under water. When we had cleared a certain amount of the muck away, we put rat traps under the suction pipes these helped greatly to prevent them choking so often.

When the suctions were cleared I went on deck and stripped the pump to clean all the valves and once again it was under way, going full bore. By the time the pump was O.K. it was about midnight and I thought I would try and get a couple of hours 'shut eye'. As I went for'd to my mess I looked into the foc'le to see how Captain Oates and Doctor Atkinson were faring with the ponies. They were having a very bad time indeed, two of the ponies were already down.

Captain Oates asked if I could give them a hand to put canvas slings under them to try and keep them on their legs. This was no easy matter as there was little or no room to move and the ponies on either side nipped us with their teeth whenever they got the chance. However, we got them slung but it was of no avail as they died before the morning. After slinging the ponies I went to my mess, what a state it was in. There was more than a foot of water on the deck swishing from side to side with the roll of the ship carrying with it articles of clothing, boots, magazines, tins of food and goodness knows what.

I had not had any food for many hours so I decided to make myself a cup of cocoa. It was a bit of a struggle getting the dry powder into the mug before going to the galley to make it. There was plenty of hot water, the chef had taken a lot of trouble to make all the kettles secure and there was a good fire. He had lashed the storm bars to the range and then lashed the kettles to the storm bars with spun yarn. Whilst I was freeing one of the kettles a great sea came through the galley door, bringing a bag of coal with it. It washed me right through the galley and put the fire out – I never saw the mug again. I picked myself up from under the fore rigging and made my way back to the mess. I gave up all idea of cocoa.

My bunk was a lower bunk, about eighteen inches from the deck and close to the ship's side. As the ship rolled the water swishing about the mess flooded it, my mattress was soaking, but worse still so was all my winter clothing which was stowed underneath it, I didn't have a dry thread of clothing in the ship. I was already drenched to the skin and had been for more than two days, however, nothing could be done about it and I was so tired I just turned in my bunk as I was, oilskin, s'wester and sea boots.

I had scarcely turned in when Mick Crean and two or three others came down and shook me, 'Come on Chippy,' they said, 'The Captain wants you to come on deck at once. The hand pump is choked again.' They had some trouble in rousing me, 'Come on Chips,' said Mick, 'The ship's sinking.' I'm afraid by this time I almost felt past caring. I had been on the go without sleep for nearly three days.

At last they got me on deck. I was wet, cold and miserable as I commenced to strip the pump assisted by the watch and scientist who were manning the pump. As I undid the nuts and bolts I handed it piece by piece to a scientist or seamen who were clinging to anything they could lay hands on; to one a top brass, to another a pump plunger and so on until I got at the foot valves which were hoisted out with a handy-billy, cleaned and replaced. Then the plunger was replaced and that section made secure. This was done to each of the other three in turn, all in total darkness with the sea breaking over everything the whole time. In the midst of the job a particularly heavy sea came aboard and washed me into the Scuppers. I could hear Mick shouting 'Get hold of Chippy' as I was washed from side to side with the roll of the ship.

Fortunately the pump did not give any further trouble for an hour or two and gradually the water in the bilge was reduced. By the morning the weather began to moderate and that evening we were again under sail and steam.

The following day we were able to take stock of the damage and losses. About twenty feet of the bulwarks washed away, two dead ponies, one dog washed overboard and a few tons of coal. Everything in the ship was sodden but things were not so bad as they might have been.

The dogs hated the wet, and looked very miserable during the gale, but they soon recovered and at dawn and dusk howled their hymn of hate as their forebears, the wolves, had done from time immemorial. There was some difficulty in getting the dead ponies out of the foc'le, for when they were put into the stalls they were intended to remain there until we reached South. Somehow they were dragged through the skylight over the foc'le deck and dumped overboard, after which it was possible to clean out the stalls of the other ponies.

The seamen were getting a very rough time, they lived in very cramped quarters on the lower mess deck, really sleeping quarters, under the ponies. Each shared a hammock with his opposite number on the other watch, as one turned out the other turned in, just as they were. Every bit of their clothing in the lockers was stained with urine from the ponies, that soaked through the leaky decks as the ship groaned and strained.

Gradually we got things ship shape again. I got rid of the water in our mess by boring holes through the deck, it was the easiest way out. We managed to dry some of our dunnage in the engine room. We were lucky enough not to get any more really bad weather before reaching the ice-pack.

Everybody was cheery again and always ready for a joke. Captain Oates put the buzz around that Mick's rabbit had given birth to a litter and sent everybody to him for a young one. He promised about nineteen people they should have one when they landed. The rabbit, however, was a gentleman and was landed with the southern party, and lived for a long time in the hay for the ponies.

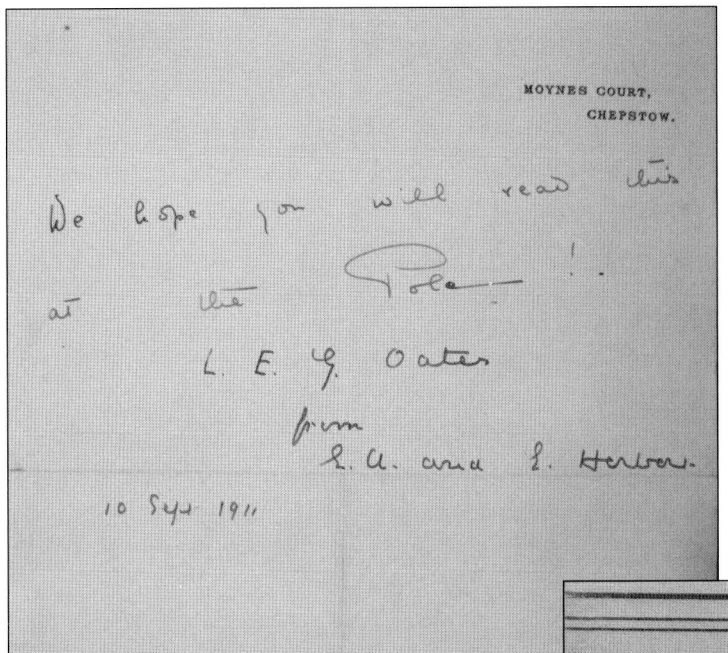

This book and accompanying letter were found in the Francis Davies collection. It is believed these items, sent to Captain Oates by Colonel Herbert, a colleague of the Inniskilling Dragoons, were left with Davies by Oates, when he departed for the Pole.

Dogs tethered aboard Terra Nova

Chapter IX. Christmas Day in the Pack-Ice

About a week after the gale we entered the pack-ice. At first it was light and fragmentary, in light streams, which gradually became heavier and heavier the floes were at times many hundreds of acres in extent and from ten to thirty feet thick, sometimes completely barring our progress.

The ship was conned from the crows nest (barrel) at the top of the main top gallant mast. From here the orders were shouted to the helmsmen as the officer tried to find a way through the floes. Often it was necessary to charge them at full speed, to try and shift or split them. This meant going ahead and astern many times to clear a space to manoeuvre the ship and get her stern pointing in the right direction. She would then go full speed ahead to gather as much way as possible. On striking the floe the ship would shiver from end to end, and her broad stern would rise, and either the floe would split or the ship just slide back again and another and perhaps many more attempts would have to be made before there was any further progress.

These measures were only taken if there were prospects of the floes becoming looser or open water ahead, otherwise we would wait until a change of wind or gale loosened the pack. A lot of patience was needed in the pack, for it is of little use trying to kick against the forces of nature, you just have to wait for it. Often we were held up for days on end, with fires banked to try and save our precious coal.

There were many large bergs almost always in sight, some of gigantic proportions. These huge masses of ice are really ice islands that break away from the ice barriers that surround the Antarctic Continent. Some of them are miles in extent, only about one eighth of their volume is above water. The largest berg seen during the Scott expedition was twenty one miles long and at that time was the largest berg ever sighted. In a later expedition we saw one ninety miles long in the lee of which we sheltered from a gale, the sea was quite calm although the gale raged overhead.

The Antarctic bergs are generally tabular in shape and stand sheer out of the water, about two hundred feet high. As they meet the warm ocean currents the under water portion slowly melts away and this upsets its stability, causing it to tilt in which position it might remain for some time, until eventually it will turn right over or split. This often happens quite suddenly sending up huge columns of water. Before the berg becomes much weathered the layers of precipitation of snow can be plainly seen.

Bergs travel with the current irrespective of the direction of the wind, possibly this is the reason why sailors say that an ice berg always travels to windward. They force their way through the pack as though it scarcely existed and leave broad lanes of open water in their wake. Most of the bergs are dead white shading to a very delicate tint of blue, except when they have turned over a time or two when they become somewhat discoloured and dirty looking, and all sorts of fantastic shapes.

Sometimes after drifting about the Antarctic for long periods it may become stranded and remain so for years, and then disappear almost suddenly during a storm or particularly high tide, or it may just disintegrate. Ice molecules appear to become fatigued in time, this seems to be the cause of disintegration.

Quite a number of them have large patches of black showing, probably lava from some volcanic upheaval countless ages ago.

Terra Nova

These have often been mistaken by ships for islands, which unaccountably disappear. In a later expedition we searched for some of these islands over a fairly wide area round about the positions reported but without finding any trace of them.

One particularly fine berg had a jet black band, about seven feet wide and almost parallel with the water line running the whole length of the berg, it looked as though it might have been painted on. Another close to had a beautiful black dome.

On two occasions I saw what looked like ice bergs on fire in the setting sun. On each occasion it was just one or many bergs, the whole berg seemed a mass of flame, I could hardly believe my eyes. I shouted to others to come and see the phenomenon. Not one of them had ever seen such a sight before. It was obviously due to the angle the setting sun made with the face of the berg.

I have counted as many as four hundred bergs of various sizes during a four hour watch in another quadrant of the Antarctic. The sky over the pack-ice is usually bright and clear, but over open water it is dark and lowering, an indication of conditions for some distance ahead. The beauty of the pack on a sunny day is indescribable with whales blowing between the floes, seals basking in the sun and maybe a few Adelie penguins would bob up on the floe close to the ship, and show tremendous interest as they came towards us.

We imagined the penguins were fond of music and often a gramophone was brought on deck and played to them. Their favourite tune, or so it appeared to us was 'God Save the King', anyhow, they didn't show the same interest in the records of the 'Dollar Princess' or 'Belle of New York'.

A few crab-eater seals, a cream coloured seal that frequent the pack were shot for scientific purpose. We usually got the livers, which were very good fried with onions, much like pig's liver and not in the least fishy. The flesh, though dark in colour was not bad eating, a little strong and oily perhaps, but one soon went off it. We also ate penguins. They were not cooked as we cook poultry, fillets were cut off the breast and fried in beef dripping, prepared this way it was quite nice. Care had to be taken to remove all the blubber from the fillets before cooking otherwise it tasted oily and rancid.

A day or two after entering the pack we came up against a large area of ice quite unlike the ice floe which had probably broken away from a low part of the Great Ice Barrier. This threatened to stop our progress southward, but from the crows nest a very narrow lead of open water was spotted some distance away. When we came up to it, it was like a canal through fairyland just wide enough for the ship to pass through comfortably, and about ten feet above the water on either side. Had it closed we should have been crushed like an egg, it was a risk but worth while.

We had managed to eke out the fresh water till we reached the pack, now we could get as much as we wanted by melting ice from the floes. When we spotted a floe on which there were blocks of fresh water ice we made fast alongside and iced ship. Fresh water ice was easily distinguishable from the salt water ice by its blue tint, salt water ice being grey in colour and flecked with salt. A party with ice-picks descended on to the floe and dug out the ice which was filled into baskets and hoisted onboard.

On the engine room casing was a tank with a steam coil. The ice was put into this tank to thaw, the water, thus obtained, ran through pipes to the fresh water stowage tanks, for'd. It was a slow business and was kept going from the ice that had been heaped on the poop.

Everybody became less fastidious in the matter of washing, once a week was often enough, usually Sundays, before Divine Service in the saloon. Sometimes one might be lucky enough to scrounge a drop of hot water from the engine room and bath on the gratings round the boiler, but as often as not would be dirtier after the bath, from the coal dust that was everywhere.

One Sunday whilst we were in the pack, Captain Oates asked me if I could tap a barrel of wine. 'Yes' I said, 'I've put taps in many barrels of beer in my time.' It was a barrel of sherry stowed in the lazarette. When I arrived to do the job Doctor Atkinson and Captain Oates were both there. I knew little about wine and expected it to be lively like beer and took the same precautions, wrapping paper around the end of the tap. I drove the tap in very carefully expecting it to squirt all over the place. After the cask had been tapped Captain Oates drew off a pint mug and said, 'The tapper always takes first drink.' I had a good swig at the mug, it was good after nearly a month of abstinence and was handing it back to him when he said, 'You haven't taken half a drink, you must finish it off.' I said, 'Not on your life', I thought they were trying to see me off. I did not know what effect a pint of sherry would have, had it been beer it would have been a different matter, I knew my capacity.

We celebrated Christmas Day held fast in the pack, alongside a very large floe. In the morning there was a service in the saloon which was decorated with the sledging flags of the Afterguard. At lunch time we had the usual Christmas fare of turkey, goose and plum pudding and very liberal 'Splicing of the Main Brace'. Christmas Day was the birthday of no less than three of the members of the expedition.

In the afternoon there was a sing song in the crew's quarters. Some of the dogs were taken on to the floes and exercised with the sledges and many of the Afterguard were on skis. In the evening there was a football match on the floe. It was pretty heavy going in soft snow wearing sea boots.

Whenever possible soundings were taken of the depth of the ocean, and the biologists took samples of the plankton. Doctor Atkinson, who was parasitologist, had the disagreeable job of examining the intestines of seals and penguins for internal parasites. We were three weeks working through the pack, about three hundred miles before reaching open water in the Ross Sea. As we left the pack we were unlucky enough to get another gale and it was thought prudent on account of the animals which were getting weaker, to return to the shelter of the pack-ice.

Soon we sighted Mount Sabine in South Victoria Land, about a hundred miles away. It was a grand sight, silent and majestic. At first I thought it was a mirage. I had heard the old timers talking so much about them.

Soon the twin peaks of Erebus and Terror, named after the ships of Sir James Clarke Ross, who discovered them, some twelve thousand feet in height showed clearly against the pale blue sky. Both are volcanoes and from Erebus a big plume of thin grey smoke drifted slowly away on the light breeze. Terror was not active. As we drew closer to Ross Island we could see the sheer face of the Great Ice Barrier stretching away to the eastward as far as the eye could see.

We were making for Cape Crozier, the extreme western end of the Barrier. The sea literally teemed with life. Many large whales were blowing lazily as they fed on the krill and schools of Killer whales, the wolves of the sea passed on their 'lawless' occasion. Hundreds of Adelie penguins could be seen swimming in the clear water, like fish, making a 'phit phit' noise as they just broke surface to breathe without in any way slackening speed.

Close to Cape Crozier the ship was hove to and a whale boat lowered for Captain Scott to get a closer view of the Cape to see if it would be possible to land and establish Winter Quarters. Here the end of the Barrier was fairly low, about twenty feet in height, but a very exposed position, even then heavy swell was breaking against the Barrier. Had it been possible to land the Southern Party there it would have saved many miles to the Pole. However, Captain Scott considered it was unsuitable for many reasons, so decided to make for McMurdo Sound on the other side of Ross Island.

As we steamed along close to the island, we could see an enormous Adelie penguin rookery stretching for miles on the lower slopes of Erebus and Terror. There were millions upon millions of penguins nesting there, and the sea was literally alive with them. Sea Leopards, that prey on the penguins were having a great time, they played with a penguin as a cat plays with a mouse, tossing it in the air and catching it again until, tiring of the fun, it gave it one shake and shook it right out of its skin, which floated away as the Leopard made short work of the carcase.

We reached the edge of the fast ice in McMurdo Sound, some fifteen miles from Hut Point, where Captain Scott had wintered on his previous expedition, and about six miles from Cape Royds where Sir Earnest Shackleton had wintered during his expedition.

About three miles distant was a low spit of land showing patches free of ice and snow. As this looked a likely place to winter the ship was made fast to the ice by ice anchors, and Captain Scott with Lieutenant Evans went over the sea ice to explore the possibilities. Captain Scott decided to establish Winter Quarters here and called it Cape Evans, after the second in command.

The work of unloading began immediately. First the dogs were got on to the ice and made fast to the hawsers holding the ship to the ice. Whilst this was going on several penguins popped up out of the water on to the sea-ice and became curious, making their way as fast as they could towards the dogs. Generally speaking they have no enemies except in the sea and came on quite fearlessly. The dogs were almost mad with excitement and strained on their chains to the limit. The penguins still came on until the leaders reached the dogs and were torn to pieces, but that did not deter the remainder, they were absolutely unable to realise what had happened to their pals and shared the same fate. Meares and Demetri had their work cut out to keep the penguins at a safe distance.

I went ashore with Captain Scott to select a site for the hut. It was quite easy to land on the Cape from the sea-ice, except for some confused ice at the tide crack, caused by the rising and falling of the tide, which was bridged over so that the sledges could drive easily on to the Cape. A site was selected on one of the bare patches of black lava, on a gentle slope that only required a small amount of levelling. The surface was like cinders, quite loose, but a few inches below it was frozen solid. This formed a good foundation for the hut.

By this time the advance party was already landing stores which included a very large tent to house those working ashore building the hut, receiving the stores and looking after the animals. When the tent had been erected, wire jackstays were laid out for tethering the ponies and dogs. I, with my party of hut builders commenced to level off the site. We worked away until midnight and by this time all the ponies and dogs were ashore. We were ready for a flying start to erect the hut in the morning. The weather was much in our favour. We had continuous sunshine which lasted, without a break, for eighty four hours. As it seemed warmer in the sunshine than in the tent I turned into my sleeping bag on a patch of snow, in the lee of a few bales of hay.

Turning in was not a complicated business, it was only a matter of removing my ski boots before getting into the sleeping bag, balaclava and all. I slept very well, in fact I overslept, and was awakened by the sledges which had started to arrive with stores from the ship. It wasn't so pleasant turning out, I felt like a chicken that had just hatched out and got very cold. I could not understand why my feet were so sore until I realised that my ski boots had frozen like iron. Fortunately they were fairly big for I was wearing three or four pairs of woollen socks.

Francis Davies working on the hut at Cape Evans

Mr Mate, who came with the first sledge was not too pleased to find us still in our bags. So we wasted little time over breakfast.

We got on like a house afire with the hut. All the framework was erected with the exception of a part which had not come to hand.

The hut was well insulated against the cold. It had three thickness of flooring with ruberoid between each thickness. The sides outside were, first a layer of matching, then a layer of Gibson seaweed quilting, and finally weather-boarding. Inside there were two layers of matching with a layer of seaweed quilting between. The roof outside had a layer of matching, a layer of seaweed quilting then another layer of matching over which was laid ruberoid and yet still another layer of matching and finally a covering of three ply ruberoid. Inside a layer of matching which formed the ceiling.

The dogs were working, pulling light loads from the start and the motor sledges were soon got going, but the ponies were in poor condition and were given a few days to recuperate. How they enjoyed rolling in the snow, after standing over five weeks in their stalls.

The dogs work in teams of eleven or thirteen, harnessed in pairs to central trace with a leader dog ahead. The leader dogs were very intelligent animals and obeyed the orders of the driver, given only by word of mouth. Often whilst sledging the stores ashore the team spotted seals on the ice and tried to make for them, but the leader dog strained all he knew to keep them on their course. They always did their work at the run.

Ponies enjoying a roll in the snow.

Demetri, the Russian boy had been a postman in Siberia. In the winter he delivered the mails regularly by dog sledge, over a distance of two hundred miles, he could not speak English when he joined the expedition but soon picked it up. On one occasion he wanted to call Tom Crean's attention to the fact that a pair of socks he had hung over the stove piping to dry were singeing, he shouted 'Mick, socks, goodbye' at the same time clasping and shaking his hands.

Whilst the dog sledge is being loaded the leader sits ahead of the team, quietly and patiently waiting the order to go. When all is ready the driver gives the order in Russian and away they go at top speed as the driver jumps on the sledge, 'tuee' stick in hand. The 'tuee' stick is an ash stick about two feet six inches long and about two and half inches at the lower end and slightly tapered at the top. In the thin end is a six inch iron spike. Its purpose is to act as a brake to prevent the dogs upsetting the sledge as they go off with a rush. It is forced into the ice between the framework of the sledge throwing up a plume of ice crystals until the dogs settle down to a steady pace. It also serves another purpose when any of the dogs misbehave – the driver must keep them well in hand.

After a few days rest the ponies joined the sledging party, taking very light loads to begin with. Some of them were already quite frisky and on several occasions ran away from their amateur drivers, shedding their loads in all directions as they made for the piquet lines ashore, with the sledge swinging behind them.

This happened on one occasion when Ponting was taking a cine picture of Skua gulls hatching. He was so interested in his work that he did not see the pony coming. It was going straight for him, the sledge making a wide sweep from side to side as the frightened animal galloped along. All hands in the vicinity shouted to warn him of the danger, but it was not until the pony was almost upon him that we were able to attract his attention, he had just time to grab his outfit and jump clear.

There was a Skua gull rookery on the Cape, day and night it was like being in a farmyard, they made an awful din, quacking like ducks.

Brownie, who was a bit of wag, had charge of one of the pony sledges and thought it would be much easier for himself if he rode the pony. We spotted him soon after he left the ship, so did Captain Oates who went out to meet him. As he passed the other sledges he put his balaclava all askew, shortened his neck and folded his arms, telling them he was Napoleon crossing the Alps. I guess Captain Oates told him he was something else when he came up to him, anyhow we all enjoyed the fun.

There were always a lot of Killer whales close to the ship whilst she was unloading. Possibly they were attracted by the animals. We had got so used to them we hardly gave them a second thought, until one day when Ponting tried to get a good picture of them. He waited near the edge of the ice, eyes glued to his camera. Suddenly several of them shot half out of the water on to the ice, whilst others came up under. They broke off the piece of ice on which he was standing and tried to shake him off. Doctor Atkinson grabbed a rifle which was always kept handy for such emergencies and kept them at bay until Ponting was able to jump on to the fast-ice, but not without losing his camera. It was a narrow squeak, after that we were not so casual when we heard them blowing, close to between the floes.

I wanted a seal skin which I had visions of getting cured. One evening tiny Abbott and I took a sledge and went in search of one. After going about four miles over the sea-ice towards Cape Royds, we saw a very large weddell, asleep. We dispatched it and skinned it on the spot, we only wanted the skin and liver.

The skin with the blubber still attached, was much heavier than we had expected and we had some difficulty in getting it on the sledge. While we were struggling with it we heard the Killers blowing quite near and lost no time in getting under way. They had either been attracted by the smell of blood or had seen our shadows up through the ice. However, we got the seal skin and the liver safely back to Winter Quarters. Later I took the skin on board, intending to flense the blubber off when I had more time, I made it fast to one of the stanchions on the poop. I never got an opportunity of flensing it and some weeks later it was washed overboard during the gale. I was not really sorry to see it go, it was too heavy for me to handle.

Four days after we landed the weather changed and we had a 'Southerly Bluster' blizzard the ship kept breaking away from the ice, so that the landing of stores had to stop until the weather moderated. I had made good progress with the hut which was so far advanced that it could have been lived in if necessary. Whilst waiting for the weather to moderate I had a lot of extra help. Fortunately I had plenty of hammers and was able to find jobs for all. Some I set to work assembling chairs and tables which had been shipped in parts to save stowage, others did necessary jobs about the camp. One party dug a cave in a glacier close to the hut for the stowage of the carcases of frozen mutton. The physicist dug another cave in the glacier to house the instruments for recording terrestrial magnetism.

The people in the ship were by no means idle on such occasions. The biologist took advantage of the break to trawl and dredge for specimens in the sound. Ponting was a splendid photographer, always busy with his cine and camera. He and Meares had been through the Russian-Japanese War with the Japanese army. Later they toured Japan and published a book containing some beautiful photographs, many coloured. He also had a fine set of lantern slides of his travels and often gave lantern lectures during the long winter nights.

Before the hut was completed the weather became very cold and work on the outside of the hut was anything but pleasant. When the hut was completed, stables had to be built for the ponies. No provision had been made for this, so material had to be raised somehow.

They were built on the north side running the whole length of the hut. The outer wall was built up of bales of Geelong compressed fodder and coal briquettes. The framework for the roof was cut from deals forming the deck in the t'ween deck of the ship. There were sufficient odds and ends of matchbaording left over from the hut to cover the roof, also enough ruberoid.

When completed the stables were very snug and particularly so when later they were drifted over with snow. They were heated by stoves specially designed by Doctor Atkinson, to burn seal blubber, Mr Meares also used the stoves for making pemmican for the dogs. Whilst putting on the stable roof the ship had to stand off as the fast-ice was breaking up and drifting out to sea. She was manoeuvring to take up another position when she struck a pinnacled rock and hung there for two or three hours and was finally rolled off by the crew running from side to side of the ship, after a good deal of weight had been shifted to the poop in order to trim her by the stern to raise her for'd where she appeared to be hanging.

I was, I'm ashamed to say, secretly hoping that she might have to stay there the winter. I did not want to go back to New Zealand in her, but wanted to winter South. A friend of mine of the Afterguard told me there was still some doubt in my case. Captain Scott wanted me to stay South and Lieutenant Pennell, who was now in command of the ship, wanted me to return. He eventually won, he said he wouldn't go back without me as I was the only person who knew anything about the hand pump. Lieutenant Bowers who was in charge of all stores, built an annexe on the south side of the hut.

Dr Edward Wilson
Chief Scientist and Zoologist

He formed a wall of cases so they could be opened from inside the annexe and filled with gravel when emptied. It was roofed with wood from the packing cases, which held the motor sledges, and then covered with canvas.

The weather became so cold that the nails stuck to my fingers and the ends of my thumbs became frostbitten. It wasn't so much a matter of low temperature as the wind, it seemed to blow all the heat away from the body, no matter how many woollies were worn the wind seemed to get right to the skin unless windproof clothing was worn overall.

There are more or less two opinions about clothing for polar exploration, one, that furs are the best, and the other, woollens covered by light, loosely fitting, windproof garments. Personally I prefer the latter. I tried wearing woollen gloves to protect my fingers, but apart from the fact that I had difficulty in picking up the fine wire nails, as often as not I nailed the glove to the job. I got over the trouble to a certain extent by making myself a glove from Burberry wind proof material, with just a finger and thumb.

Our clothing was the finest made in England and that speaks for itself. Vest and pants were extra heavy, the vest had double backs, my pyjamas more or less of the same quality with soup plate buttons – plenty of woollies, breeches of corduroy and over all a loose blouse and trousers of wind proof material. We wore woollen balaclava helmets with a wind proof helmet over when the south winds blew. When on the march reindeer fur finnesko stuffed with senna grass were worn instead of leather ski boots. The sleeping bags were of reindeer skin with an eiderdown bag inside in very cold weather.

Being small in stature, I scored in the matter of clothing, which seemed to be all of one size to fit the biggest man in the expedition. My corduroy breeches instead of buttoning just below the knee buttoned round my ankles, all so and so and pockets. On the second Sunday after we landed we had made such progress in establishing Winter Quarters that Captain Scott ordered a day of complete rest. In the morning Divine Service was held in the open close to the hut. The leader read the prayers and psalms for the day and all joined lustily in singing the well known favourite hymns, 'Onward Christian Soldiers' and 'Eternal Father Strong to Save.'

Everybody then went off in little parties to explore the vicinity. A small party of scientists went over the Barne Glacier to Cape Royds, Shackleton's winter quarters, to collect records etc. The going was pretty heavy owing to the crevassed state of the glacier, this was the only route since the sea ice had gone out beyond Cape Royds.

I went with a party on to the lower slopes of Mount Erebus, later known as the Ramps. There were enormous heaps of lava, free of ice and snow, jet black and brick red in colour. It looked just as though some giant had cleaned out a huge furnace and dumped the heaps to a preconceived plan with an outsize in wheelbarrows. We all appreciated the day off, for we had worked zealously for long hours since landing, not so much for the rest, perhaps, as for the Leader's consideration, it gave us the feeling that our efforts were meeting with his approval.

Lieutenant Pennell met me tramping over the snow without goggles and insisted on handing me his saying, 'There are lots of navigators in the expedition but only one carpenter'.

As a matter of fact I rarely wore goggles, they irritated me too much and I never had the least sign of snow blindness, while others seemed to become so constantly. I think this is partly due to the colour of the eyes, blue eyed people were more easily affected, perhaps being a Westcountryman my eyes were green anyhow, I'm told it is very painful for about three days, just as though pepper had been thrown in the eyes and they stream with water.

Meares and Oates at the blubber stove in the stables.

Almost the last thing to be unloaded from the ship was the third motor sledge, the highest powered of the three. It was hoisted on to the sea ice and a party commenced to drag it ashore. Suddenly without any warning it disappeared through the ice, that had become rotten, and sunk in a hundred fathoms of water. One of the party went through the ice with it but was rescued and taken back to the ship, by the time he reached her his clothing had frozen stiff. It wasn't possible to salvage the motor sledge owing to ice conditions, even if we had suitable equipment.

After all the stores had been landed from the ship and while the Winter Quarters was being completed she did a considerable amount of trawling and dredging. What amazed me most was the great variety of sponges collected, not the sort used domestically, and even a small octopus, usually associated with tropical seas.

Seals, penguins and seagulls within easy reach of Winter Quarters were killed and stowed in the ice cave as an addition of the fresh meat supply for men and dogs, the seal blubber being used for fuel. Early one morning, as we were about to start work, an Adelie penguin waddled into the camp like some funny little old man who had come to see what was going on. One of the seamen knocked it out with a 'tuee' stick, another addition to the fresh meat supply, and threw it on a case until convenient to put it in the ice cave. It lay on the case all day with its head hanging over the side, to all appearances quite dead.

Imagine our surprise when some ten hours later it got slowly up and shook itself, with no more concern than if it had just awakened from sleep. It stood on the box for sometime, apparently very interested in everything going on, then it waddled off into the tent where it stood watching operations for hours on end. Nobody had the heart to 'kill' it again and it became the Winter Quarters mascot, until in due time it migrated. During the weeks it remained with us it went to sea daily but always returned to the tent and never seemed to associate with the hundreds of other penguins always swimming about.

The setting up of Winter Quarters was practically complete in less than three weeks, and the Southern Party had moved into the hut, which was very snug and comfortable. It was divided into two compartments, one for the officers and scientists, the other for the men, by a partition of cases of liquids, liable to damage by frost.

Captain Scott had a cabin to himself and Lieutenant Evans and Doctor Wilson shared another. For the most part the remainder of the officers and scientists built themselves cubicles, fitted with bunks according to taste. There was also a dark room and a small laboratory for the physicists. In the men's quarters there was a large cooking range and in the officers' quarters a heating stove.

In London, Captain Scott had stressed the necessity of good ventilation to the hut. I had a couple of ventilators made to my own design, so arranged that the piping of the range passed through the centre of one, and the piping of the heating stove through the other, as this would help to extract the foul air. These were very efficient, so much so that when a southerly 'bluster' was blowing it caused such a vacuum in the hut that papers were sucked off the table and the vents had to be blocked up.

Preparations were now being made for the first depot laying journey. It was intended to lay depots of food and equipment for men and animals as far as the seas would permit, already it was a month past the southern midsummer, in preparation for the journey to the Pole the following year. By the time everything was ready for a start on the depot laying journey, fast ice had broken out to four miles beyond Cape Evans, towards Hut Point.

As it was not possible for the sledges to reach the fast ice from Cape Evans, the party with sledges, dogs and equipment were taken on board the ship and transferred to a glacier tongue at the edge of the fast ice. The ponies were, with some difficulty, led round the glacier tongue by way of the shore, under the ice cliffs. The loaded sledges were got on to the glacier tongue and the party made ready to leave. The pony sledges were first away after good byes and good wishes all round. A party from the ship saw them safely off the glacier tongue on to the sea ice. We gave them a rousing cheer as they eventually got under way, single line ahead, led by Captain Oates – there were eight ponies in all.

Captain Scott, Doctor Wilson, Meares and Demetri left some time later with the two dog teams which was the usual routine as the dogs travel so much faster than the ponies and much better when there is something ahead of them. When the ponies were little more than dots on the sea ice, the dog teams, headed by Captain Scott, started off at a terrific pace in spite of the 'tuee' stick which was sending up a great plume of ice crystals as he tried to check them. In less time than it takes to tell the sledge had upset and rolled over and over and ended in a free-for- all dog fight.

It appeared that 'Osmond' the best leader dog, and recognised by all the other dogs as their 'king' had misbehaved himself in some way or other. For this lapse on his part he had been deposed from the leadership of the team and was one of the first pair behind the new leader. This was not noticed by 'Osmond' until the team started off. He couldn't understand why he wasn't ahead and pulled like fury to get into his rightful place. At last he got hold of the new leader and the fight became general. It was no easy matter to separate them when once their fighting instincts were roused. However, no great harm resulted, and as soon as the sledge had been righted and 'Osmond' reinstated as leader they were off again, cheered on their ways by the ship's company. When we last saw them they were tiny specks on the ice, nearing Hut Point, the first stage of their journey.

Little did we dream that we of the ship would never again look on the brave, happy faces of Captain Scott, Doctor Wilson, Lieutenant Bowers and Captain Oates, though of course, all realized the hazards of the under-taking. Taff Evans was not with this party, he went with another party unfamiliar with sledging, because of his experience.

We found the remains of one of Shackleton's depots on the glacier tongue containing a few tins of compressed pony food. We took in a good supply of ice for fresh water from the glacier tongue before proceeding to Butter Point on the other side of McMurdo Sound, to land a party to make a geological survey in the Western Mountains.

On arriving at Butter Point we landed the party and saw them well under way. We laid out a depot of food here. As we were making for the spot selected, I spotted a piece of ragged bunting fluttering in the wind on a short pole. This we found marked the spot where Shackleton had previously made a depot. It contained a few odd pieces of clothing, tins of food and a tube of Haglecream with 'David' scratched on it, these we took as souvenirs. We made our depot at the same spot. The ship returned to Cape Evans to take on board two ponies for the Eastern party and then sailed for Kind Edwards VII Land, about three hundred miles to the eastward of Ross Island.

We steamed along close to the Great Ice Barrier surveying it as we went. Near the Bay of Whales we ran into the southerly gale and had to stand well off on account of the drift (snow) blowing off the barrier causing bad visibility. On reaching Kind Edward VII land we found it impossible to land the party on account of inaccessibility. The ship then headed for McMurdo Sound and we completed the survey of the barrier.

We entered the Bay of Whales one night, about midnight, and were greatly surprised to see a ship made fast to the ice. She was Amundsen's ship, *Fram*.

We made fast close to her, Amundsen was not on aboard, but at his winter quarters, which had been established about a mile away on the Great Ice Barrier. During the fore noon he arrived by dog sledge and visits were exchanged. He had a splendid team of dogs that looked like chows, particularly well trained and driven fan-wise. He blew a whistle and the team stopped instantly.

He got off the sledge, of Eskimo type, turned it over and left the dogs contentedly sitting on the ice awaiting his return, they made no attempt to move. I don't know if they were a better type of dog for sledging than ours, they certainly looked the part, no doubt they were a very special team for the Leader. Our dogs were rather nondescript, hardly two alike, but they had chests like ponies and were very strong. Altogether Amundsen had about a hundred and thirty dogs, they were mainly fed on dried fish.

Fram was already a famous ship in Polar history, she had been with Nansen on his 'Farthest North' expedition and had since been fitted with an internal combustion engine. She was a very handy little ship and in fair weather it was only necessary for one man to be on watch as the engine could be controlled from the bridge. Amundsen originally intended to continue his work in the Arctic and I saw many cases of stores when I visited *Fram* marked for North Pole Expedition. Captain Scott had received a cable from Amundsen when *Fram* reached Madeira, to the effect that he had changed his plans and was coming South.

We hardly expected him to establish himself almost next door to the British, who had so far done all the exploration work in this quadrant of the Antarctic. Captain Scott had first broken the trails as far as the foot of the Beardmore Glacier and Shackleton, who was then one of Scott's party, had continued the work during his own expedition later, and had found a way up the mountain range which he named the Beardmore Glacier, on to the polar ice-cap and got to within one hundred miles of the Pole.

However, Captain Scott did not change his plans one scrap and by no means made a race of it. Naturally he hoped to be the first to reach the Pole, for the honour of our Country, and who was more entitled. Captain Scott did say that had he known Amundsen was set on coming into that quadrant he would have chosen another, in order to add to the knowledge of the Antarctic.

The Bay of Whales, I gathered from conversation with an admiral who served with the first expedition, was discovered by Captain Scott and named Balloon Inlet. It was here that a survey balloon was sent up from *Discovery* – later it was named 'Little America' by Byrd.

We shot the trawl before leaving the Bay of Whales and got some very interesting specimens, one in particular the biologist was very pleased about, looked like a rubber model of an outsize tulip, before the petals open, with a stem about three feet long and ivory in colour.

Adelie penguin

Commander Evans (centre) chatting with Lieutenants Campbell and Pennell

Chapter X. New Land Sighted

On return to McMurdo Sound we dropped anchor off Cape Evans and again landed the two ponies. This time they had to be dropped into the sea and towed ashore astern of the whale boat.

The first one struck out manfully and seemed none the worse for the cold bath when it landed on the beach, but the second one just gave up and floated astern of the boat and would have drowned if its head had not been held out of the water by it's halter. Sailors don't know much about horses generally but as soon as the pony got on the beach Lieutenant Rennick drenched it with a bottle of whisky and it was kept on the move to restore circulation. The treatment may have been somewhat unorthodox but most effective.

After we had landed the ponies we steamed towards Hut Point until we reached the fast ice, now several miles beyond Glacier Tongue. There were a number of large whales, Blue and Fin, and a great number of Killer whales, some with dorsal fins two or three feet in height, like lateen sails in the Sound. Lieutenant Campbell went on skis to Hut Point to leave letters for Captain Scott reporting he was unable to land at King Edward VII Land and that he intended to winter at Cape Adare at Borchgrevink's old winter quarters and that Amundsen was wintering in the Bay of Whales.

Lieutenant Campbell had hoped to find somebody at Hut Point who could carry his reports to Captain Scott but in this he was disappointed.

As soon as Lieutenant Campbell returned from Hut Point the ship went alongside Glacier Tongue and took in a good supply of ice for fresh water, this had to last until we reached New Zealand. After icing ship we returned to Cape Evans and then proceeded northward to land Campbell and his party.

The ship was very light and the coal was getting very low. Some of the pig iron ballast had been landed at Lyttleton in order to make more room for stores. To compensate for this it was intended to ballast with rock, but we only managed to get about twenty five tons aboard. For some time the ashes from the furnace had been put into the main hold, unknown to me, and this caused more trouble later with the hand pump.

Almost as soon as we cleared McMurdo Sound the weather became bad and for days we did not get a sight of the sun to fix the ship's position. As we were near the South Magnetic Pole the horizontal directive force was almost negligible, rendering the compass needle sluggish and practically useless. When the gale moderated and we were able to get sights, we set course direct for Cape Adare. It was the last chance the party would have to land, otherwise they would have to go back in the ship to New Zealand.

Cape Adare was fairly free of ice, but there was heavy surf breaking on the beach which was littered with large ice boulders thrown up by the recent gale. The ship lay off about a mile while Lieutenant Campbell and I went ashore in the pram to explore the position. We found the surf too heavy to run the pram on the beach so I backed her in and Lieutenant Campbell waited an opportunity to jump ashore. I then lay off and waited for him.

The Cape was a low spit of land not much above sea level, possibly about half a square mile in extent upon which thousands upon thousands of Adelie penguins were nesting. At the back of the Cape the land rose sheer for over a thousand feet and was very mountainous beyond.

From the pram I could see Borchgrevink's [* note 3] two small huts, one appeared to be intact but the roof had blown away from the other.

When Lieutenant Campbell returned to the beach I backed the pram in and he jumped for it. Just as we pushed off I noticed the pram was making a lot of water and we then discovered that the plug had been knocked out when she bumped on the beach. As we couldn't find it Campbell stuffed his handkerchief in the plug hole and held it there while I pulled with all my might for the ship. When we arrived alongside he jumped out and left me to tie up the pram as best I could, not thinking, perhaps. I felt nearly at the last gasp and hardly had strength to pass the painter aboard, all the time she was rapidly filling.

Campbell decided to winter there and the work of unloading began at once as the season was getting very late and the pack-ice might come in and trap the ship, this had to be avoided at all costs. I went ashore with the first boat load of stores, together with my assistant hut builders, Tiny Abbott and Brownie. After unloading the boat we levelled off the site for the hut. The place fairly reeked of guano from the penguin rookery. I did not envy those who had to spend so many months there, perhaps when it became snow covered it did not smell so badly.

As we dug we came across the remains of countless numbers of seals in all stages of decomposition and their skeletons. It must have been their graveyard since the beginning of time. It took some time to get the material for the hut ashore or, at least sufficient to enable us to get a move on, using only two small boats. In the meantime we scouted around to see what we could find. There were scores of cases lying about outside the huts that had become weathered and bleached with no marking to indicate their contents. I stuck a pick into one case and found it was ball ammunition, luckily I did not strike the 'business' end of the cartridges or it would have been 'Bob's your uncle'. There was a solitary dead dog still chained to the wire jackstay. Inside the hut things were a bit untidy, as though the previous occupants had left in a hurry.

Snow Goggles

There was a range and plenty of coal so we soon had a nice fire going and made ourselves cocoa au lait, several tins of which we found there.

The pack-ice came in on the ship during the afternoon and she had to put out to sea. However, we now had sufficient material to allow us to get on with the hut. The ice went out again on the turn of the tide and the ship returned to land the remainder of the stores. In about thirty hours all the stores for the party had been landed, the framework of the hut had been completed and covered with a layer of match boarding.

Owing to ice conditions and the lateness of the season the Captain of the ship was anxious to leave as soon as possible. It was therefore decided that the party would have to complete the hut themselves, in any case there was Borchgrevink's hut which, though small, could be used in the meantime.

The ship now had about thirty tons of coal in her bunkers. This meant that we would have to sail most of the way back to New Zealand. In order to do this it was necessary to make westing, so we steamed westward skirting the edge of the pack. Two days after leaving Cape Adare we sighted new land. What a thrill it gave us to think we were the first humans ever to gaze on it. We could not get within several miles, owing to the pack but as far as possible made a running survey. This was named Oates Land and the main physical features were named after members of the expedition, one of the bays was called after myself, Davies Bay [* note 4].

We were hoping to find an open lead of water so that we could get closer, instead the pack closed in on us. We thought we were trapped for the winter and took account of our resources. However, two days later the wind changed and the pack loosened so we were able to get clear to the northward. We were now down to our minimum reserve of coal, twenty five tons this we would require as the ship approached New Zealand so fires were let die out. Almost as soon as we were clear of the pack the gales commenced, even before we had time to get things lashed and secured.

I remember being called early one morning to put the weatherboards on the saloon side lights, which were near the deck, I had just got one of them out of its stowage when the ship gave a tremendous roll. My feet shot out from under me and I slid across the poop on my behind, acting as a sort of human squeegee for the slush and half frozen sea water, still hanging on to the weatherboard, a cumbersome thing, about six feet long and eighteen inches wide. I slithered past the helmsman, first to one side and then to the other, who was of course unable to help me, being more than fully occupied with the wheel. As I slid to leeward I could look right over the rail, which was buried in the sea, it seemed I must go right overboard but as she rolled back it just saved me. Somehow, I don't know how, I managed to pick myself up still hanging on to the weatherboard. By the time I got them shipped I was wet through to the skin and half frozen for I had not yet donned my sea gear – sea boots and oilskins.

Day in and day out for nearly three weeks we had nothing but gales, westerly gales. The ship rolled to an unbelievable angle, so much so that in passing along the deck one often got a foothold on the side of the hatch combing whilst clinging to the life line.

The hand pump too, on which the safety of the ship, in her leaky condition, depended so much was giving a lot of trouble. All hands, when not pulling on ropes, were manning the pump. Since landing the shore parties the seamen were reduced to seven in a watch, but they were real seamen, sailing ship men, we shipped several of them in New Zealand, they could take on any job at sea, and were almost as much at home in the stokehold as they were on the top gallant yard.

The Captain, Lieutenant Pennell, set a splendid example to everybody. He never undressed the whole of the voyage and slept in a sleeping bag when he did sleep, for he only had cat naps, mostly by day, on the shelf in the little shelter on the bridge protected from the elements by only a canvas curtain. There was one particularly bad night when we were reduced to main lower topsail and reefed fore sail. The gale roared like thunder through the rigging without the least sign of a lull, drowning all the sounds. Most of the hands were aloft furling the half frozen canvas, whilst a few of us tended the ropes on deck. I won't say to what degree the ship was rolling for it would not be believed.

How the hands clung on to the yards at all I don't know, yet they did, and never lost a stitch of canvas.

To add to the eeriness there was a very vivid display of the Aurora Australis which was even reflected in the drops of water hanging on to the yards. It had one advantage, it gave sufficient light to see what we were doing.

The gale continued with unabated fury all the next day. Lieutenant Pennell and I went into the main hold to inspect the ballast, he thought from the behaviour of the ship that it had shifted. What a scene of desolation the practically empty hold presented in the dim light of a candle lantern, with the bilge water breaking over the rock ballast like seas on a rocky shore. However, it had not shifted, there was so little of it when spread over the hold I don't think it could have made much difference if it had. The Captain thought we ought to have a sea anchor handy in case the weather got even worse. Normally we should have used a spare spar on the end of a hawser but all these had been landed at Lyttleton.

I made one with the help of the seamen, from the 'tween deck planks which were bolted together and then covered with two or three thicknesses of old hatch tarpaulins, and further strengthened with steel wire ropes. On one corner was slung a heavy ice anchor to keep it upright in the water. We did not use it, however, as we could not afford to drift before the wind on account of the necessity to make westing or we might be unable to make New Zealand. The alternative was to run before the Westerlies and make some port in South America, but we were in no shape for such a long voyage as there was no great quantity of food stores remaining. Everything that could possibly be spared had been landed for the shore parties, and rightly so. We used oil which was allowed to perculate through oakum in the for'd and after W.C. pans but I could not see that it made much, if any difference in preventing the seas breaking.

The hand pump suctions became so completely choked I was unable to clear them in the ordinary way, and the water in the hold rose rapidly so that it swished from side to side up to the main deck beams on either side of the ship as she rolled. There was only one thing for it, the pipes would have to come down to be cleared. This was not so easy as it sounds as they were of cast iron and had probably not been unbolted since the ship was built, apart from this the bolts were in a very cramped position so I had to use a hammer and cold chisel on them and being cold weather the pipes might easily have fractured. However, the risk had to be taken, but before starting on the job I explained the position and what was likely to happen to the Captain. Alf the Bos'n volunteered to give me a hand with the job.

As we disappeared into the trunk one of the officers asked anxiously, 'What are you going to do?' I told him it was a case of to Hell, or to Lyttleton, and hoped for the best. Once inside the trunk the hatch was battened down to keep out the seas. Alf slung one of the pipes, with a rope to secure it, after the bolts were out. I then performed on the nuts. Unexpectedly the nuts started fairly easily and we soon had the lower section of one of the suction pipes, about twelve feet long, off.

It was a risky business with the heavy cast pipe swinging about in the trunk like the pendulum of a clock, but we managed to get it cleared by using iron rods and soon had it rebolted again. I tapped on the hatch to let them know on deck that the pump could be started and then we took down the lower section of the other suction pipe, got it cleared and rebolted. Once again the water came full bore, through the wide mouth of the 'Old man of the sea', as it discharged on deck. That evening the Captain sent for me and said; 'A few more grey hairs I suppose Davies, but you have saved the ship once more.' I replied I did not think it was as serious as that. 'There's no need for mock modesty', he said. At the same time he told me he was writing to Captain Scott to bring the facts to his notice, Captain Scott was destined never to receive this letter.

There were many days, after the Captain had fixed the noon position of the ship, when we would find the distance still to go to New Zealand many miles more than the previous day even though we appeared to have made a good day's run but westing had to be made.

The chef, now single handed, had a tough job preparing the meals but nothing was ever too much trouble for him, he took a real pride in his work. He was in his galley before six mornings and would still be found there after eight in the evening with, perhaps, a couple of hours 'shut eye' in the afternoons. He was very popular with the crew who would do anything for him, at any time. He always kept the old stock pot going, lashed to the range, and everybody helped themselves to a cup of hot soup during the night watches. Only once was he unable to provide a hot meal and that was one breakfast time, during the worst of the gales, even then he struggled aft to the lazarette to get slabs of chocolate which he served out to all hands as they manned the pump.

We were never without fresh bread, I generally gave him a hand with the breadmaking, after the parties had been landed South. When the dough was ready I weighed it and he put it into the tins ready for the oven. We had to keep both galley doors shut to prevent being hove out on deck when the ship rolled. When it was necessary to turn the bread in the oven I would get hold of Walt [* note 5]round the middle, as he stooped to the oven, with one arm and hang on to the handle of the galley pump with my other hand, it was more than a single handed job to keep the oven door open and turn the tins.

In very bad weather it was a bit of a struggle to get the food from the galley, and even more so to hang on to it while eating. On these occasions the chef and I teamed up. In our little bug hutch of a mess there was a corner between the food cupboard and the bunks into which one could just squeeze. Into this corner we fitted an empty milk box to sit on, then we took it in turns to have our food, whilst one fed the other hung on to the dishes or teapot. Sometimes, somebody would forget to lash the cupboard doors, then they would roll open and out would come tins of milk, jam, mugs, cutlery and what-not and fall straight into the chef's bunk. Often he had to collect all this dunnage before he could turn in and as often as not scrape strawberry jam or milk off his blankets.

At last the day dawned when we could safely raise steam again for the last lap of our journey. What a relief it was when we sighted The Snares, a few rocky islets to the south of New Zealand. We had visions of a real night's rest. It had been very wearying being thrown all over the place, day in and day out, even in our bunks we had no real rest, we had to sleep with one eye open hanging on to the leeboard the whole time to prevent being thrown out on deck.

We dropped anchor in one of the bays of Stewart Island where a member of the cable company was waiting to receive Captain Scott's dispatch for the newspapers. To comply with the terms of the news contract nobody was allowed ashore for twenty four hours so we sailed within an hour or two of Lyttleton.

Often I have asked myself why the old Salts endure such hardships, voyage after voyage, year after year for most of their lives and still come up for more. The only answer I can give is comradeship, brotherhood of the sea if you like, the faith they have in each other, no matter the danger, for death is always just round the corner in a sailing ship in the Southern Ocean, with its perpetual gales and countless ice bergs, creates a sense of security, strange though this may sound.

I still cherish the hope that someday, somehow, I shall see the Southern Ocean again. Maybe I shall have to wait until I become an Albatross, sailors say they are the souls of old shellbacks.

Thomas Clissold ship's cook, never without fresh bread

Chapter XI. Surveying the Three Kings

At Lyttleton we were glad of a rest and were given a few days leave. My pal, the chef, and I had been recommended to spend our holiday at Akaroa on the Banks peninsula, a pleasure resort not a great distance from Lyttleton, about eight hours by rail and coach. We went by rail via Christchurch to a small township called Little River, the terminus of the railway in that direction. From here we travelled by stage coach and four, we felt we had stepped back into the middle ages. Akaroa is situated on the shores of a beautiful land locked bay with high hills all round, obviously the crater of an extinct volcano.

It was here that the British landed and hoisted the Union Jack when they took possession of New Zealand, only three days before the French arrived. Many of the French immigrants settled there in a part of the Bay called French Farm which looks like a bit of France, with its lines of poplars. It is a very steep climb from Little River, up the winding road cut in the side of the hills, before reaching the hill top, the lip of the crater, where the traveller by coach gets his first view of the Bay.

At the hilltop there is a hotel called, aptly enough, The Hill Top Hotel, where we stopped for lunch and fed and watered the horses. It was a lovely drive from the hilltop down to the little township nestling on the shores of the bay. The surrounding hills, once covered with bush, had been cleared and were now excellent growing land. Hundreds of charred tree stumps still littered the slopes. Akaroa is known for its dairy products, butter and cheese, particularly the latter, and it also produces a quantity of grass seed for export. Many of the seamen on the coast take a holiday reaping the grass seed. There was a modern creamery, even in those days, and it was a never to be forgotten sight to see the churns of milk in huge wagons drawn by teams of oxen, arriving in the early mornings.

We put up at the Maderia Hotel. It was the sort of place one dreams of, comfortable, good food and an atmosphere of peace. Hardly had we arrived before several young people from the other hotels came along and invited us to a picnic, by motor launch the following day. We had a wonderful time cruising around the bay, anchoring at times to fish. When we tired of fishing we landed for games or rambled through the bush. We always went ashore for lunch and boiled the 'Billy', boy scout fashion. All the hotels catered for that kind of holiday and supplied excellent luncheon baskets. The launch was owned by a Greek by the name of Dominic. He and his wife were a nice old couple and often of an evening we would go to their home for a sing song.

We had picnics every day. I don't remember the weather being anything but perfect the whole time we were there. How we appreciated the peaceful life after the endless rolling of the ship. It was the best holiday I have ever had. I wonder in these ever changing days whether it is so peaceful there now.

A penguin that had somehow become separated from its kind, took up its abode in the township. It became very tame and was known to all as 'Pompey'. It used to wander about the place almost like one of the humans, though I never saw it carrying a shopping basket or queuing for fish.

After our holiday we got busy refitting the ship. She was dry docked to see what damage she had sustained when she struck the rock off Cape Evans. A part of her keel had been torn off and her garboards almost penetrated. Few ships would have stood up to that ordeal. I had certain alterations made to the pump suctions which did away with a good deal of the trouble we had previously experienced. When the defects had been made good the ship was painted from truck to keel. She looked a grand old lady of the sea, as pretty as her figurehead, a Goddess scantily clad with one hand to her breast and the other extended as though to ward off or calm the sea.

As it was some months before we were due to sail again for the South we were fortunate in getting a job from the New Zealand Government to survey a group of islands called the Three Kings, north of Cape North, North Island. These islands were uninhabited, the only sign of civilization being a small hut for shipwrecked mariners, stored with cases of biscuits, milk, clothing, ammunition, gun, fish-hooks etc. It was visited regularly by the New Zealand Government steamer *Hinemoa*.

We were told that some eleven years earlier a ship on passage between Auckland and Sydney, Australia, was wrecked on one of the Islands with heavy loss of life and at the subsequent Inquiry the Captain maintained that the Islands were wrongly charted. This, I believe, from the result of our survey had some foundation in fact.

Before sailing we had another spot of leave which the chef and I spent at New Brighton on the coast about six miles from Christchurch and easily reached by tram. Some of our New Zealand friends, members of the Union Rowing Club, Christchurch, lent us a small seaside bungalow, called a 'bach' in New Zealand. It had only two rooms, a kitchen with a cooking range, the other had four bunks fitted on two sides of it and a piano, it was in fact drawing room come bedroom come dining room. It was located on a quarter acre section, together with a boathouse, where several pair oared and four oared racing boats were housed. At the bottom of the section was the river Avon, the beach was about four hundred yards away.

We provided our own food and did our own cooking which, generally speaking, only amounted to breakfast, we got most of our meals at the local, the New Brighton Hotel. Early mornings, before breakfast, we had a turn on the river in the pair oared boats which wore off the fog of the night before and gave us an appetite for breakfast. In the forenoons we visited the local for a few beers and usually had a game of billiards, Walt was a fairly good player. In the afternoon we bathed in the sea and sunbathed on the sand dunes. It was a splendid beach extending for some fifty miles and at low tide there was a very wide stretch of fine sand ideal for all beach sport and bicycles, there were few motor cars in those days.

In the evening we usually met some of the rowing club boys at the local and occasionally at the 'bach', generally weekends, when we had a pretty hectic time. We were all very pleased to get to sea again. Like the Chancellor we had a certain amount of difficulty in balancing our budget, we found the gap between income and expenditure an ever widening one, we always had a dead horse to work off.

To us the survey was more like a yachting cruise. At night we anchored in some sheltered bay close to the job, usually on one side or the other of North Cape, according to the wind. When at anchor we spent a lot of time fishing and had some fine sport often catching two or three large fish at a time, it seemed almost a matter of the number of hooks on the line. When they were feeding we caught snapper as fast as we could haul them in. They were very good eating and when we caught more than we required for our immediate use we wind dried them at the masthead, where the flies could not reach them, this move we picked up from the New Zealand 'bush whackers' who always hung their fresh meat high up in a tree. According to them when the flies are about to blow their eggs they are too heavy to fly very high.

About once a month we visited the nearest settlement, Mongonui, for the weekend. It was a bit of a one horsed sort of place, with the usual store that sold nearly everything, including beer. Apart from the store there were less than half a dozen shacks.

We always kept a look out for the steamers running between New Zealand and Australia. Some of them got to know us very well and often threw a case or two of fruit overboard to be picked up by us. I believe the passengers enjoyed the thrill of seeing our boys dive overboard and swim for the cases as much as we enjoyed the fruit. Anyhow it was less trouble than lowering a boat. Occasionally we had an afternoon off and landed on the beaches for recreation. I remember one of the beaches where we bathed in a fresh water lagoon on one side of the pebble ridge and in the sea on the other.

There were great heaps of sand here from which we dug large pieces of volcanic glass, dark green in colour. We also found a skull possibly of some sailor, shipwrecked or buried by a passing ship, more likely still the skull of some Maori chief. According to Maori folklore the Maori brought their Chiefs to the north of the island for burial where they could see the sun rise out of the sea. Old Jock, one of the seamen, was very perturbed when the skull was brought on board, he, like most seamen, was very superstitious and predicted all kinds of disasters. On top of a small islet I found a very large bone which I thought was part of some prehistoric monster. I was disappointed when the biologist told me it was part of the vertebra of a large whale, I've seen thousands of them since.

I was of a party of three landed on the extreme north of North Island to take a round of angles by theodolite. It was a pretty barren spot except for tea tree scrub, wild flax and large ferns. Ages and ages ago this area was covered by a forest of giant kauri pie trees from which the sap had run into the ground.

Digging for the sap, or Kauri gum, as it is known commercially, is one of the minor industries of New Zealand. It is much like amber, usually unclear, and often pieces are found containing prehistoric insects or small plants, which had become enveloped by the sap. The ground was full of holes where the gum diggers, mostly Australians had been at work.

On our way back to the ship a small herd of wild pig came racing in our direction. I had heard a good deal about the wild boar and stood by to plunge into the small lake, along the edge of which we were walking. However, they passed by without even deigning to notice us, some were white, others ginger and a few piebald. They were descendants of the pigs liberated by Captain Cook when he discovered New Zealand, and lived chiefly on fern roots.

When we reached the beach to pick up the boat it was almost dark. A large school of Snapper was feeding close in shore in very shallow water, so shallow that their dorsal fins were right out of the water. They were feeding on the small mussels with which the rocks were thickly covered.

On another occasion I landed with Lieutenant Rennick and seaman, Bill Bailey, on the largest island of the Three Kings, to erect a beacon on the highest peak for shooting angles, we waited several days for suitable weather. The weather was fine with a smooth sea and long oily swell when we lowered the boat in preparation for the 'expedition'. Hardly had it touched the water when the engineer started the circulating pump, the discharge overboard of which was right under the whale boat davits, half filling the boat with water and giving most of us a good sousing by way of a good start. On reaching the shore at what we thought the best place to land, we found the swell breaking heavily on the rocks and it was with difficulty that we managed to land with all our gear. The boat shoved off as quickly as possible and returned to the ship.

Kauri gum, collected by Francis Davies, much like amber it is the fossilized resin of kauri trees, forests of which once covered a great deal of the North Island of New Zealand

In spite of the fact that this looked the easiest way up, the cliff was practically unscaleable, at least for the first twenty feet. However, by standing on each others' shoulders the topmost man Rennick was able to claw his way up the rock until he reached the tea scrub, after that it was comparatively easy. We threw him a rope and he hauled up all the gear, I being lighter than the seaman, stood on his shoulders and was hauled up by Rennick, then we hauled up Bailey.

We first made for the hut for shipwrecked mariners, on a neck of land between two of the peaks. Here we deposited everything but the theodolite and axes for clearing away the bush that grew thickly all over the island.

It was a stiff climb, well over a thousand feet up the hill. When we reached the summit we found the highest peak was the hill beyond. When we eventually reached the highest peak we cut down the scrub to give a clear all round view of the flag staff.

By this time it commenced to rain very heavily and we were soon drenched to the skin, for we were only wearing five bob dungaree bush-whackers suits. The climb had taken us much longer than we expected and it was getting towards evening when we started on our way back to the hut, with the job still uncompleted.

As we made our way back we saw literally millions of Mutton birds, black sea birds somewhat smaller than a rook, streaming in from the sea to roost in burrows in the ground. The Maoris eat them. It is said they taste like mutton, hence the name. It was dark long before we reached the hut and we had a good deal of trouble making headway through the low scrub. Once Rennick, who was leading, suddenly disappeared and we thought he had gone over the cliff until he assured us he was all right and hanging on to a branch of a tree. Fortunately we had some matches in a watertight tin. When we struck a light we found his feet were only a few inches off the ground, he had fallen into a sort of fissure or water course.

We reached the hut at last, drenched to the skin, cold and miserable but we soon had a good fire going on the blind side of the hut, so that it could not be seen by the lighthouse keeper across the strait who might mistake us for castaways. We boiled the 'Billy' for tea and had a good meal of biscuits and tinned sheep's tongue. Most of the night we sat around the fire drying our clothes, there was plenty of dead wood lying about to keep the fire going. Between times we tried to get a nap on the biscuit boxes in the hut, but we were too wet and cold to sleep and soon returned to the warmth of the fire. Rennick produced a flask of brandy which helped to keep the blood circulating.

The following morning it had ceased raining. What a transformation, the birds were singing and the sun shining gloriously, everything around us except for the hut was just as it had been from the beginning of time. On the cliffs in the distance there were a few goats which had been put on the island for castaways. We hoped to get one or two, fresh meat for the ship. We had little time to admire the scenery and as soon as we had had a meal we returned to complete the beacon. About midday we were surprised to hear one of the ship's company hailing us from the distance. It appeared that nothing had been seen of us since we landed and the Captain had become a little anxious.

We had completed the job by this time and were on our way back. We were pleased to hear from the seamen who had hailed us that the boat was laying off our landing place. When we reached there the swell was too heavy for the boat to come alongside, so after passing the instruments on board by line, we had to swim for it.

We had a gale during the survey, it lasted for nearly a fortnight, but we were quite comfortable, hove-to with a sea anchor out and the spanker set. We had plenty of sea room towards Australia.

Whilst hove-to a school of fish came around the ship. The officers got out their rods and the seamen a long bamboo with a hook and line on the end which they took in turns to work up and down. Strange to relate the experts with the rods did not get a fish, though they had many bites. There was great cheering every time the seamen caught one, and they caught quite a few with their crude outfit. The fish were either an unknown species or one with which the biologist was not familiar. All were preserved for scientific purposes.

On our way to Lyttleton, after completing the survey, we called at Mongonui on purpose to get some varnish for the steering wheel, which Rennick had scraped and sandpapered to make the ship look 'tiddley' when she reached harbour. He gave me a sovereign and sent me ashore to see if the store had any, whether they had no varnish or whether I had forgotten about it I don't quite remember, as I had not returned after about four hours Rennick sent Alf to look for me.

By the time we thought it proper to return to the ship we had varnish nor quid, but were in merry mood. He wasn't very pleased when I told him I had been unable to get any varnish and another quid was chalked up to the 'dead' horse. It was a lovely carefree life and if at times we hankered for the bright lights, well, we were broke anyway, we always were.

At Lyttleton we were given a couple of weeks leave to recuperate after the four months surveying. Walt and I, not being over flush financially, decided we would spend our leave quietly at the 'bach', New Brighton. Our 'dog-robbers' (civvy suits) were also beginning to show signs of extreme old age. However, we got over the latter difficulty fairly easily. We talked nicely to Bill the steward, and he unearthed a couple of suits belonging to some of the party South. One of the suits might have been made for me, it fitted me perfectly. Walt was inclined to be a little corpulent, the waistcoat of the other suit would not meet by inches and the trousers, permanently turned up, were about two inches too long.

I set to work to make the necessary alteration. I opened up the back seam of the waistcoat and put in a gusset. The only material I could find was a piece of green Burberry from one of the tents. It didn't exactly match the grey lining of the waistcoat but that was of little concern when he had the coat on. I took a reef in the bottoms of the trousers giving them a double turn up. To further increase our wardrobe we bought some plain buttons to replace the 'gold' buttons on our uniform suits, so they could be worn as civvies.

Walter had a portmanteau big enough to hold all our dunnage, with room to spare. Two ground ash walking sticks completed our going away outfit. We intended to travel in uniform but I had no cap so I sailed aft to the saloon where I saw one hanging on the peg. I tried it on, it was just my fit, with a couple of newspapers inside the sweatband.

The following morning we set off by rail for Christchurrch. We reached there about elevenish, full of the joys of Spring, firmly resolved to go easy on the beer in order to stretch the finances. Before catching the bus for New Brighton we had to do a bit of shopping for groceries and meat. Everything in the garden seemed to be going lovely until we spotted what looked to be a nice, quiet pub.

I looked at Walt, he looked at me and we both said, 'What about a quiet one?' nearly knocking each other over in our haste to get in before we changed our minds. Lummy, a nice quiet pub! Most of the ship's company must have had the same happy thought, for nearly all of them were there. A great cheer went up as we entered, bang went our good resolutions. Two pints please. After a few rounds of beer, chiefie (the chief engineer) suggested we should get ourselves straw hats as the weather was very hot, so we tossed for them the loser to pay for three.

I was the 'victim'. We went to a shop just across the road and asked for three straw hats. When we were fitted I asked the assistant how much. He said, 'Seven and sixpence each.' I said, 'You don't happen to have anything a little more expensive I suppose.' He said he was sorry he hadn't. 'Thank Heavens for that.' I said, or words to that effect.

It was late in the afternoon when we realized we would have to go if we were to get our shopping done before the shops closed. We got the groceries all right, they were wrapped in two paper parcels, also two dozen eggs in cartons of a dozen. Then we made for the tram shelter, carrying the portmanteau between us on the two walking sticks, and each had a parcel and carton of eggs under one arm. On the way down to New Brighton there was a little argument with the tram conductor about smoking and ended by our names being taken, we 'rumbled' he intended to have us before the beak.

When we got out of the tram we made for the 'bach', in line abreast. What with the portmanteau between us and making heavy weather, the side walk was hardly wide enough for us. Walt had the outside berth, walking in the gutter, that had the nice cool stream of water about six inches deep flowing through it. We had not far to go before we reached the sand dunes which were planted with lupins, then in full bloom, to keep the sand from blowing away. I tucked Walter under the lupins, leaving him with the parcels of groceries and cartons of eggs, whilst I shouldered the portmanteau and took it to the 'bach' some quarter of a mile further along.

When I returned for him he was fast asleep using one of the parcels for a pillow, this had burst and the sugar had run into the sand. After a little coaxing I got him under way and we proceeded under easy steam and in due course reached the 'bach'.

I could not find the key of the place, it was not in its usual place under the wall plate. Walter shoved me out of the way to look for it himself and brought me to earth on top of the eggs. At last he managed to find the key and after getting the gear inside I put him into one of the bunks and tied a wet handkerchief around his head. I sort of felt sorry for him imagining myself to be all right. I left Walt fast asleep and made my way to the local, where I was rescued later by some of the rowing club boys.

The next morning we tried to piece together the events of the previous day, our uniforms told the sad story of the eggs. We remembered the incident in the tram and wonder if we could do anything about it. We decided our best course of action was to see one of our friends, a New Brighton councillor and ask him to use his influence on our behalf, this must have worked for we heard nothing about it. It took me a whole afternoon sponging egg and sand from our uniforms but we had no flat iron to press them.

I scouted around to see if I could borrow one from the neighbours but unfortunately we didn't know any of them personally. There were some children playing in the distance so I went along and asked one of the small boys if his mother had a flat iron. He said. 'Yes, Mister.' 'Ask her if she would kindly lend it to me.' In a few minutes he brought it along. Soon I had the uniforms pressed and hanging out on the clothes line to air off, looking brand new. I changed the buttons at the same time to turn them into 'civvies.' By lunch time we had recovered somewhat from our alcoholic depression. In the afternoon we set off for the beach to bathe and brown ourselves off, again full of good resolutions. In the sand there were millions of shell fish, about the size of cockles and similar in flavour, called pippies. We picked a large handkerchief full.

We were very pleased with ourselves, thinking this Heaven sent supply of free food would materially help to reduce our 'house-keeping' expenses. In the evening we cooked the pippies and had quite a blow out. What thirst they created, it torpedoed all our good resolutions and the evening found us, as usual in the local.

One afternoon we met Alf on the beach with one of his friends, a widow. Alf was almost a native of Christchurch, he had been in Lyttleton so many times with previous Antarctic expeditions and had a wide circle of friends who, strangely enough, seemed nearly all widows. He told us they were on their way to visit some friends, another widow and her daughter, and said he would be pleased if we would join them. We said we would be delighted, it was all the same to Walt and I, we didn't mind trying anything once. We had a very pleasant musical afternoon, the widow was a good pianist and her daughter a clever impersonator.

The daughter gave one or two impersonations of the 'Handcuff King' who had recently performed in Christchurch, by the faces she made I thought she was also a good contortionist. Alf sang several sea shanties including his old favourite 'She had A dark And Roving Eye,' and Walter, who had a good voice, sang several old ballads, I played the audience.

Rather late in the afternoon a cup of tea and a damp biscuit was handed around. Walter had not drunk much of his tea when he drew my attention to a ring around the inside of the cup, from previous use. I was very amused at his discomfiture until I found a similar ring in my cup. Our hostess suggested to her daughter, for our hearing, 'perhaps the young gentlemen would like to stay to dinner.' But with the cups still in mind we said we deeply regretted we were unable to accept on account of a prior engagement.

Whilst at the 'bach' this time, we had an invitation from the members of the Union Rowing Club to a water carnival, to mark the opening of the rowing season. I had helped to rig up one of the boats for the carnival, as a motor car, a motor car was almost a novelty then, for one or two of the boys and their girl friends.

On the day of the carnival Walt and I decided to walk the seven miles to Christchurch by the less frequented roads, in order to see a bit of the country. We again resolved to give the beer a miss until we reached Christchurch, and passed the local in grand style. After walking about a couple of miles, along an unfamiliar road we came upon a hotel we had never heard of. This somewhat caught us aback for we thought we knew all the 'dives' for miles around.

We felt it almost a 'duty' to give it the once over, particularly as it was a hot day and we had already walked two miles on a dusty road. It was lovely and cool in the hotel lounge and the beer was good. After two or three pints we settled down to steady drinking.

Walter suggested a game of billiards. As he warmed up to the game he took off his coat, he was wearing the waistcoat with the green gusset. By this time there were quite a few spectators. I whispered to him, 'You are wearing your billiard waistcoat.' He lost no time in putting on his coat. The hotel had a nice friendly atmosphere, or so we thought as we took our departure late in the evening. Needless to say we did not see the carnival.

The New Zealand pubs, or hotels as they prefer to call them were divided into two classes of bars, the lounge and the 'stiffs' bar. The drinks in both bars were the same, excepting the price. In the lounge beer was sixpence and in the other bar threepence. If you wanted spirits you were handed the bottle to help yourself. On a side table in the lounge there was always cheese and salad, and hot rolls at lunch time, to which you helped yourself, free. In the 'stiffs' bar there were generally cold roast potatoes and hot roast potatoes, or rolls, at lunch time. We usually did very well, the barmaids always took a special interest in our welfare.

One of the dogs on the Terra Nova Expedition

Chapter XII. Preparing for the Second Voyage South

We had now to prepare for the second voyage South.

The ship was in better shape than for the first voyage and the living quarters made a little more comfortable. A small heating stove had been fitted in our mess to dry up some of the dampness. This made it cosy even if everything did get very sooty and the chimney had to come down every two or three days, for sweeping.

Captain Oates told me, as he was about to leave for the depot laying journey, that the Indian government was presenting seven mules to the expedition and asked me to look out for them. We were also taking more dogs and a lean-to shed, of simple construction, as an annexe to the Southern Party's hut. In due course the mules and dogs arrived and were quartered on Quail Island until the ship was ready to sail. Stables were built for the mules on either side of the fore hatch, between the galley and the ice-house, forming a good shelter from the seas.

The mules were a cross between the ass and the hardy Tibetan pony, no two were alike in colour. The Indian Army Transport had taken a lot of trouble training them on the snow slopes of the Himalayas and had actually put them on rockers a few hours a day, to get them used to the motion of the ship. A New Zealand friend of Lieutenant Pennell's volunteered to look after them on the voyage South. I took eight wild rabbits, alive, in a hutch lashed to the top of the stable, hoping they would make a nice pie for the shore party.

The main hold was, as usual, filled with coal, the fore holds and 'tween decks with relief stores and the ice-house with its full complement of frozen mutton, many carcases also festooned the mizzen rigging.

We did not sail until 15 December, nearly a fortnight later in the season than the previous year. In order, if possible, to avoid being held up so long in the pack-ice. The weather was not so bad on this voyage. We had gales of course, but they were not nearly so bad as the one that almost brought disaster to the expedition before it had scarcely begun.

Lieutenant Rennick had two hens in a coop over the saloon. He could not understand why they never laid any eggs. I think Alf the Bos'n could have told him; it was a case of the early bird. Alf kept the morning watch from four to eight a.m. He used to send a seaman to see if the hens had laid before he went off watch.

The dogs caused a bit of inconvenience for they always used the ropes coiled on deck to do their business, it was more comfortable for them than the wet deck, but it was not so pleasant in the darkness to pick up a rope so fouled, especially as washing water was scarce.

The weather being fine the Skipper decided to get the mules out on deck for a little exercise, and at the same time clean out the stables. Everything went according to plan and all had been put back into their stalls, with the exception of one, the most difficult, that had to be backed in. Lal-Khan – for that was the mule's name, they all had Indian names, appeared to be a comparatively quiet animal and for this reason had been left until last. The Skipper, who was playing the leading part, brought Lal-Khan along and tried very carefully to humour him through the narrow opening into the stall. He would not have it at any price, he rose on his hind legs, kicked and jumped about on top of the forehatch in the small space between the stables which also housed the for'd winch.

Davies with one of the seven mules presented by the Indian Government which were brought to shore base when the Terra Nova returned to the Antarctic in February 1913

Two or three others gave the Skipper a hand but only succeeded in exciting the mule more. Then the Skipper had an idea. He took out his handkerchief, a red one, and blindfolded him. Then as many hands as could get to the mule pushed him back into his stall by man force. These mules appeared to be more intelligent than horses, they were up to all sorts of tricks and could kick in any direction, there was a distinct leer on the their faces as they got one home.

At night times when the ship was more or less quiet, they would stamp on the deck of their stalls and make an awful din. I'm quite convinced they knew it prevented the watch below getting any sleep. The ponies we had on deck on the first voyage South had metal feeding boxes hung on the front of their stalls. When we got into the really cold weather their tongues stuck to the metal and became very sore. Captain Oates pointed this out to me when he told me we were having mules and suggested wooden boxes, so I had some made in Lyttleton. Every morning the wretched animals had gnawed their feed boxes till there was little more than the iron brackets left and I had to make others from empty cases before they could be fed. They looked at me as much as to say, 'What do you think of that, chum?' I coated the boxes liberally with Stockholm tar to try and discourage them, they liked them better that way.

For some reason or other, somebody had let go the lashings of my rabbit hutch on top of the stable and it rolled off on to the deck, letting the rabbits loose. Two of them mistaking the scuppers for rabbit holes ran into them and right through into the sea, then there were only six.

There was some excitement on Christmas Eve. The cat fell overboard. One of the dogs on deck made a grab for him, he jumped for the main rigging but missed it and went clean overboard. The lifeboat was called away. The ship was making seven knots at the time. The cat soon fell astern, everybody kept a sharp look out for him. Many albatrosses and other sea birds circled just over him. We were afraid they might pick his eyes out, for according to the old shellbacks the Albatross picks out the eyes of anyone unfortunate enough to fall overboard. When the boat came up to him he was striking out manfully. In twenty minutes from the time he fell overboard he was back on board again; not a bad effort considering the boat was well lashed and had to be cleared away before she could be lowered. There was a heavy sea running at the time but there was no lack of volunteers to man the boat.

On Christmas Day we had a southerly gale with squalls and heavy snow, the seas breaking on board the whole of the day, we sighted many icebergs the first for the voyage. In spite of the weather we carried on with the season's festivities. At nine thirty in the fore noon there was a service in the saloon, we sang some of the old Christmas hymns including 'Noel' and 'Good King Wenceslas.' After church we were served out with two pint bottles of beer, fifty cigarettes, twenty-five cigars, oranges, apples, nuts and sweets.

I gave my rabbits a treat for Christmas, some of the mule's food and fresh carrots. We had turkey, plum pudding and mince pies for the midday meal.

In the afternoon I relieved the helmsman so that he could join with the rest of the boys in a sing song under the foc'le. My trick lasted for nearly three hours, and I received a great cheer when I returned to the foc'le after being relieved which I acknowledged with a 'graceful' bow. Everybody was merry and bright. The Warrant Officers, which included myself, dined with the Captain and Officers. We had a jolly evening, many songs were sung accompanied by Lieutenant Bruce and the second engineer on their banjoes.

Many large bergs were sighted during Boxing Day. In the dog watches we saw the ice blink in the sky and met the pack before midnight. It extended to the eastward as far as the eye could see, but was only two or three miles across and was of small heavy floes that had been subjected to heavy pressure. Before reaching the pack we sounded in a depth of 1801 fathoms. There were a number of crab eater seals and penguins on the floes and many whales blowing in the open lead of water.

For two or three days the pack was fairly rotten but became heavier and we had to bank fires and wait for it to open up, which it did two days later. Trimming coal from the main hold to the bunkers on the poop was a heavy and constant job, and we had not a great number of hands, so it was a case of roping in every spare hand.

Magnetic observations were taken on the floes, well away from the ship. Cape Adare was sighted at midnight on the 3 January, but owing to ice conditions the ship was unable to approach until six p.m; we were pleased to see the flag flying from the hut. She hove-to about a mile and half from the shore. I was detailed by the Captain to go ashore in the first boat. It was not easy to land as there was very heavy surf breaking on the beach, littered with fragments of ice that threatened to hole the boat.

The party, all well, were on the beach to greet us, what excitement and handshakes. We immediately started to transfer stores and specimens to the ship but after only one boat load the ship had to put to sea to avoid pack-ice. I had to stay ashore all night. What stories we had to tell each other. They, of course had not heard anything from the outside world for over a year. All their mail had been sorted out and tied up in separate pillowcases.

They had made a very good job of the hut and had Borchgrevink's until it was completed. When they moved in they had a lot of trouble with the cooking range, it burnt very badly. All the seamen had a go at it in turn and tempers became a bit frayed. Finally Brownie took the thing to pieces and laid bare the source of all the trouble. The spare parts, cheeks, fire bars, scrapers and what not had been stowed in the back of the ranges to save space and to keep the parts from becoming separated from it. This was really a good idea, but they were unable to see it in that light.

Another snag with the range was, it had to be stoked every five hours and as it had to be kept going twenty fours of the day somebody had to get up during the night to stoke up. They took it in turns, one night each. The night watchman always read the meteorological instruments.

This brought to light another snag, they did not have an alarm clock, so the leader called for suggestions. The doctor had a brainwave. The general idea was to find out how far a candle burnt down in five hours and then to bore holes in other candles at that distance and pull a piece of twine through, so that when it burnt down to the hole it burnt the twine and released a weight which dropped into a tin bath. This was very effective, not only did it wake the watchman but everybody else, it made enough noise to wake the dead. Brownie improved on this. He used the same idea up to a point but instead of tying the twine to a weight he tied it to a thin piece of cane that acted as a spring. On the cane was a loop attached to the starting handle of the gramophone. When the twine burnt through, the cane flicked the lever and started the gramophone. The favourite record was one of Caruso.

During the long winter night Brownie framed two or three pictures from odd pieces of moulding left over from the hut. These were hanging on the wall over his bunk. He called my attention to them asking me what I thought of his handiwork. I said, it looked a pretty good job for an amateur. He had mitred the corners and then tied pieces of wide braid over them. Tiny Abbott and Dickason joined in to tell me how clever Brownie had become with the tools and what a fine job he had made of the frames.

Then Brownie pulled off the braid and exposed the mitres joined more like a 'seizure' than a fit and they all laughed like pixies. They had pulled my leg good and proper.

The penguins migrated during the winter and returned in the spring. When they began to lay, two buckets full of eggs were collected daily, and stored in a lean-to shed on the end of the hut, a natural refrigerator. These eggs were intended for the ship but unfortunately we were unable to get many of them aboard. The Adelie penguin eggs are about the size of domestic duck eggs and taste exactly like them, there is no fishiness of flavour.

Many of the penguins, close to the hut, whose nests had been robbed sat on empty milk tins or such like, one tried to cover half a hemispherical cheese tin which was at least eight inches in diameter. They seemed just as contented as they would have been sitting on their eggs, anyhow, they would not have to feed these later on.

When the ship returned she managed to get within a mile of the beach and the transhipping of stores proceeded rapidly. Then she hoisted the recall as the ice was closing in on her and we had to leave in a hurry after securely fastening the hut.

On leaving Cape Adare we made for Robertson's Bay, to land the Northern Party for a month's survey in the vicinity. Sledges were being loaded, when it was discovered that in the hurry to get away the party had forgotten the sledgemeter, an instrument trailed behind the sledge to record the distance travelled. Lieutenant Campbell asked if I could make one, I said 'I thought I could' and got on with the job. The only instrument I had for recording distance was the 'clock' of a sounding machine, which registered in fathoms. Anyhow, the engineers and myself got over the job, they made the metal fittings and I made the wheel from three ply etc. Campbell told me later, that it worked perfectly. In a way the wooden wheel was a godsend, they smoked it in lieu of tobacco during the time they were forced to winter in the glacier – but that is another story. We gave them a hand to land all their paraphernalia, they had a month's sledging stores and emergency stores for another month, we were to pick them up at the end of a month.

There was very heavy pack all along the coast but the ship forced her way to the fast ice in Robertson's Bay, we called this particular bay, Terra Nova Bay, which was about a mile and half from the shore.

From Terra Nova Bay we made for Granite Harbour, to pick up a geological survey party from Cape Evans. On our way we passed the Drygalski Glacier that juts out into the sea for several miles, it was a grand sight. It is actually afloat and rises and falls with the tides, but has scooped out such a hole in the seabed that it is held almost as in the dock.

We were unable to get within miles of Granite Harbour, though we tried again and again to force a way through the heavy pack which clung to the coast of Victoria Land, possibly due to easterly winds, neither were we able to get nearer than thirty miles to Cape Evans.

In cold weather the 'Seamen's heads' (W.Cs) became frozen solid and it was most disagreeable job keeping them clear. Several of the seamen asked me to rig up a temporary affair on the foc'le, pointing out that unless something was done about it, sooner or later somebody would meet with a serious accident, they didn't put it exactly like that but that was what they meant.

Judged by the standards then existent in sailing ships, the 'heads' were fairly modern, at least they must have been an improvement on those in Noah's Ark for they had an iron trough and twin seat, double banked as the sailors say, and were vastly superior to a New Zealand 'Thunder Box' as there was an unlimited supply of water for cleansing purposes. Indeed there were the seven seas to work on, it was only necessary to dip it from over the ship's side by means of a draw bucket, kept handy and made fast to the pin rail. At times people might 'forget' to use the draw bucket, in bad weather or if nobody was about to 'remind' them, not intentionally of course, oh dear no! It was a jolly little excursion using the draw bucket in a gale, and the ship rolling thirty or forty degrees, as one clung for dear life and tried to avoid the swinging door taking off fingers or breaking legs.

With the help of some of the seamen I lashed two spars on to the foc'le head, as near the jib-boom as possible, protruding over the side sufficient to allow the twin W.C. seat to be a nailed to them, and so that the occupants could hang on to the rails to prevent the possibility of falling into the sea. Overall I fitted a framework which was covered with a tarpaulin as a certain amount of protection from the elements. The front and bottom were wide open and it was necessary to climb over the rails to use the contraption. Unlike the man who fell out of the balloon and had all the world before him they had all the world behind them.

It was one of those rare fine days in the Antarctic, very bright sunshine and not a breath of wind when it opened for 'business' I received congratulations galore, I had even to mention as discreetly as possible in order not to 'hurt' their feelings, that I preferred not to have my statue in Trafalgar Square. But what difference the weather can make to one's creature comfort.

A day or two later we had a real snorter, with heavy snow right up the hawse pipes. How they cursed me as they passed my bench on their way to the foc'le head, holding their unbraced trousers with one hand so as not to waste time. I replied by asking them what sort of explorers they called themselves and gave them a little pep back-chat on what was expected of such heroic fellows and offered them a magazine to while away the time.

I must admit that with a head sea it was not too pleasant as she dipped her bows and sent showers of cold spray over the foc'le. On these occasions, too, there was always trouble with the 'Bumph' it refused to go down and after tickling one's ear would perform all kinds of aerobatics as it made its way aft, towering, side slipping, waving to the officer of the watch on the bridge and, as often as not crash landing on the deck. Lieutenant Rennick fixed up a broom handle with a nail in the end with which he collected these offensive missiles. On one of his routine rounds he said to me, with a twinkle in his eye, 'Lots of telegrams about today, Davies.' I replied, 'Yes, Sir, and they all appear to have been autographed.'

It was not until 3 February that we were able to make contact with Cape Evans. In the meantime we were kept very busy battling with the pack-ice trying to relieve the Granite Harbour party; trawling and dredging for specimens, taking magnetic observations, surveying the coast as far as possible, exercising the dogs on the floes and endlessly trimming coal.

Terra Nova in the ice

Chapter XIII. The Polar Party Sets Off

About midday on 3 February we anchored to the fast ice two miles south west of Cape Royds, and seven miles from Cape Evans. At six p.m. we saw two dog teams coming over the ice from Cape Evans, Doctor Atkinson and Demetri with one team and Mr Meares and Mr Simpson with the other. They were all looking very bronzed and fit. Doctor Atkinson with three others had just returned from the top of the Beardmore Glacier, his was next to the last supporting party.

He reported the Polar Party was in good condition and very hopeful of reaching the Pole. Eight went forward towards the Pole but about half the party, the last supporting party, would turn back at about one hundred and twenty miles from the pole. The ponies and dogs went as far as the foot of the Beardmore Glacier, from there onward the sledges were man hauled, about three hundred and forty miles to the Pole and about seven hundred and fifty back. The ponies had done splendidly. The last five days to the foot of the glacier was very heavy going in deep snow, and there was no food for the three remaining which were then shot. The other ponies had been shot as the loads became reduced and were fed to the dogs.

It was estimated that if the Polar Party only did twelve miles a day they should return to Cape Evans between 10 and 15 March on full rations. No ship had previously remained in McMurdo Sound so late in the season and returned the same year. They also reported that no traces of Amundsen had been seen and that all Shackleton's depots had been found.

They had very bad luck on the first depot laying journey (1911). Shortly after departure from Cape Evans the sea ice broke to the south of the Cape and severed communication between the party and the station. The eight ponies and two dog teams were occupied till 30 January establishing a base camp on the ice barrier seven miles from Hut Point. Twenty seven miles further south another depot was established, named Corner Camp. The snow surface proved very soft, making the work terribly hard for the ponies, and a three day heavy blizzard was a further severe trial to animals in poor condition.

On 8 February they preceded south marching by night and resting by day in order to aid the animals. The weather was exceptionally bad but fortunately the surface improved. The three weakest ponies were sent back, but these were caught in another bad blizzard and two died. The remaining ponies and dogs reached latitude 79½ deg South on 16 February where a depot was made and the party then returned. More than a ton of stores were depoted at One Ton Camp, as it was named.

On the way back the whole of Captain Scott's dog team fell into a crevasse. Captain Scott and Mr Meares were caught on part of the snow ridge over the crevasse along which they were travelling. Most of the dogs hung by their harnesses and were only freed with great difficulty after three hours. One of them was badly injured and afterwards died. On his return to Hut Point, Captain Scott found the dispatches concerning *Terra Nova* and Amundsen.

More stores were sledged to Corner Camp. The party then returned to Hut Point. Captain Scott and Captain Oates remained at Hut Point to try and save a pony which had been badly hit by the blizzard. Lieutenant Bowers, Mr Cherry Gerrard and Mick Crean, with four of the best ponies, set out to follow the dogs but they had not gone far when they encountered working cracks in the sea ice so they turned back marching for four miles.

In the middle of the night the tired ponies obliged the party to camp. Early in the morning Bowers was awakened by a noise and found that the ice had broken all round the camp and was moving with a heavy swell. One pony had disappeared from its picketing line, and was never seen again.

Hastily packing their sledges, the party tried to work their way over the pack, dragging the sledges and jumping the ponies from floe to floe. About noon the party reached the Barrier, but found it unscaleable. The swell was churning up the ice and breaking heavy floes against it. Mick Crean volunteered to try and obtain help. He travelled over the moving pack to find a break in the barrier and eventually got on to the barrier by wedging his ski stick in a crack.

Owing to the weather conditions Captain Scott became anxious about the party on the sea ice, and with Doctor Wilson went on to the Barrier. Doctor Wilson, through glasses, saw the ponies adrift on the sea ice. Later Crean was seen approaching. Captain Scott, Captain Oates and Crean set off immediately and by good fortune discovered the missing party.

By means of an Alpine rope the men were rescued with difficulty. Then they succeeded in salvaging their sledges and their loads, but could do nothing for the ponies, which were only thirty yards away. They left the ponies with full nosebags and took a rest. Later the pack became again stationery.

The party worked their way along the Barrier, and finding the ponies made desperate efforts to save them. Bowers and Oates risked a long detour over the pack and jumped the animals from floe to floe, whilst the others dug a trench through the lower part of the Barrier edges. The floes were high above the water and very uneven, Killer whales were blowing all round between the floes. Only one pony won through, the others failed to jump and were lost.

Discovery hut, a light Australian type of building, was found almost completely filled with hard snow. Windows were broken and the door unhinged, but was cleared and repaired and afforded good shelter. The western geological party returned, bringing the total number at the hut to sixteen. Huge land, ice falls on the slopes of Mount Erebus prevented any possibility of returning to Cape Evans by land, but with the bays freezing an attempt was made to reach the station, partly by land and partly over the sea ice. The party was caught in a storm on the sea ice, but reached Cape Evans in two days. All parties had returned to Cape Evans by the end of May.

In the middle of the winter Doctor Wilson, Lieutenant Bowers and Mr Cherry Gerrard started on a sledge journey to Cape Crozier to observe the incubation of the emperor penguins at their rookery. Very heavy going forced the party to relay their sledges for the main part of the trip, and the outwards journey took a fortnight. The temperature was seldom above minus 60 deg F and often below minus 70 deg F, the lowest being minus 77 deg F.

Behind a ridge of land on the slopes of Mount Terror they spent three days building a stone hut, which they roofed with canvas. They had great difficulty in crossing the huge barrier pressure ridges in the dim noon twilight to reach the rookery, but were successful on the second attempt. Comparatively few birds were found on the rookery, but these had begun to lay even at this early date. Some eggs at different stages of development were collected which should give considerable information concerning this interesting bird.

That night they had a violent gale, from the force of which they had poor shelter. Gusts of hurricane force blew down on the igloo, and a tent and other carefully secured articles were blown away. The canvas roof of the hut was torn to ribbons and for more than thirty hours they were confined to their frozen sleeping bags, half buried in snow and rock debris. It was forty eight hours before the wind decreased and they were able to get a meal.

Figurehead

Terra Nova

A search down wind was made for the lost articles. They were fortunate in finding the tent caught in some rocks practically undamaged. They were forced to turn homeward, but on the return journey they were held up for two days by another storm. The party returned after five weeks absence, encased in ice and suffering from loss of sleep, but otherwise well. It was the first winter journey in the Antarctic and was a great feat of endurance.

The southern journey started later than was originally intended, in order not to expose the remaining ponies to the great cold of early spring. On 22 October the first party started on the Polar journey, Lieutenant Evans, Day, Lashley and Hooper, with two motor sledges. The motors experienced unexpected difficulty on the sea ice where it was thickly covered with snow. They went a little better on the Barrier, but were finally abandoned about sixty miles on the Barrier. The party then man hauled the sledge until they were overtaken by the ponies.

Captain Scott left Hut Point on 2 November marching by night and resting by day in order to give the ponies the benefit of the warmer day temperature. He reached Corner Camp on 9 November. They followed the tracks of the motors until they found them abandoned. They were delayed by a blizzard but reached One Ton depot on 16 November. The dog teams had caught up some days earlier and the whole party now proceeded in company. The animals were given a day's rest at One Ton depot.

The ponies did remarkably well. The first were shot for expediency but could have travelled further. The animals had ten pounds of oats and three pounds of oil cake daily and travelled about fifteen miles with a load of about six hundred and fifty pounds. At every halt snow walls were built to protect the ponies.

For a part of the journey they experienced very bad weather. The wind was very violent and at times a great amount of snow fell. The ponies and tents had to be continually dug out. Then the temperature rose to plus 35 deg F and the snow melting on the equipment soaked everything in water.

The going was very heavy, the leading pony wore snowshoes and the men hauled on skis. It took fourteen hours without a meal to do eight miles. Doctor Atkinson wanted to take me ashore to Winter Quarters as there was a lot of work for me to do but the Captain would not allow me to go in case I should be cut off from the ship.

The physicist saw the ship miraged in the sky a fortnight earlier, first the right way up and then upside down. We must have been at least thirty to forty miles from Winter Quarters. We started to ballast the ship with rocks from the shore two miles away, but only got about three hundred weight, the distance was too far.

As there was every appearance that the sea ice was definitely breaking up and the ship would in all probability be close to Cape Evans within a matter of hours, the Captain allowed me to go over the sea ice to see what jobs there were to do. The season was already getting late and the ship would have to leave in about a month or become frozen in for the winter. I still had the annexe to the hut and stables to build for the seven mules.

Lieutenant Rennick, Tom Williamson and myself left the ship on skis, each taking two dogs on the chain, of course. We thought they would be a great help to tow us along. Unfortunately, our Russian vocabulary was very limited though we made up for it in nautical language, but the dogs took not the slightest notice, they went in any direction except straight ahead. The surface of the ice was very rough with little snow. It was the worst possible surface for skis which constantly became scissored throwing us heavily, it was then the dogs pulled their hardest.

After going about three miles we came to a lead of open water, about thirty feet across. Whilst I sat on the ice holding the six dogs, Rennick and Williamson looked about for a piece of ice, that would be used as a raft to ferry us across the lead.

I had only been hanging on to the dogs a few minutes when two or three penguins bobbed up out of the lead. The dogs were after them in a second, dragging me with them, I dare not let go or we would never have seen them again. I lashed them with my ski stick but it had no effect. Fortunately the other two were between me and the penguins and managed to bring them to a halt. It was a good thing I had a strong seat to my pants. The penguins were driven off and we ferried ourselves across the lead. Once across we chained the dogs to our waist and carried our skies. When our dogs heard the dogs ashore there was no holding them. I arrived at full speed, full length on the ice, skis and all. Some of the shore party caught them before they reached the other dogs and there was a free for all.

The people at Winter Quarters were out of the hut to greet us and there was much handshaking. They all looked very fit but many of them, who formed part of the supporting parties, showed the scars of badly frostbitten faces with nasty sore looking split cheeks, noses and lips, which looked as though they had been scalded. There was, of course, a lot to tell and a lot to hear on both sides, everybody seemed to be talking twenty to the dozen. It was like a Portuguese grog shop, everybody talking and nobody listening. I returned to the ship in the evening and it was not until two days later, 6 February to be precise – that I packed my tools and gear on a dog sledge and went ashore to get on with their work. The ship was still three miles from Winter Quarters. I was made very comfortable in the hut, Patsy Keohene and I yarned until it was almost time to turn out in the morning. He had a wee drop in the bottle.

The ship started discharging, landing the mules first. They were tethered on a patch of soft snow. How they enjoyed rolling in the snow after seven weeks aboard ship. After the mules came the material for the annexe and stables. What with landing the stores, and Doctor Atkinson and Demetri preparing sledges to meet the returning Polar Party, I got very little assistance. Patsy Keohane, and the cook, helped between times, and a few others as they could be spared.

All that remained of the stables built the previous year was the roof, most of the coal and forage that formed the outer walls had been used. By the evening I was hung up for material but had completed three of the seven stalls.

After turning in I heard more stories of how the party had fared during the long winter nights, some funny and others not so funny. During the winter Doctor Atkinson got lost in a blizzard within two hundred yards of the hut. After walking about for six and half hours trying to find Winter Quarters, he came up against a steep black cliff this was Razor Back Island about two miles away from the Cape, in the sound. He dug himself in and waited for the weather to clear. By this time he had been missed and search parties in threes, burning flares and firing guns set out to look for him. He neither saw nor heard any of them, but during a short break in the blizzard the moon shone and he was able to see the direction of the hut. He returned without seeing any of the search parties. One of his hands was very badly frostbitten and swelled to a tremendous size and took a long time to heal.

Lieutenant Evans, Gran and Forde sledged more stores to Corner Camp in the early spring. One of Forde's hands was so badly frostbitten that gangrene set in and he had to be transferred to the ship and Williamson landed in his place.

In two days the stables were completed and the mules moved in. Everybody said how comfortable they were and how pleased Captain Oates would be when he saw them. A week after coming ashore I had practically finished my work at Winter Quarters. The annexe gave them another three hundred square feet of room for stores, and would save going out into the snow to dig them out as they had to before. It was bitterly cold working on the annexe and my fingers, cheeks, nose and lips became split with frostbite. A day or two after the annexe was completed there was a heavy blizzard that completely drifted it up, so that it was possible to walk right on to the roof. The drift is as fine as flour and finds its way through almost invisible cracks.

During the blizzard the ship was blown out of McMurdo Sound and went on to pick up Campbell and his party, but was unable to get within thirty miles of Terra Nova Bay owing to very heavy pack.

Doctor Atkinson and Demetri left after the blizzard to meet the returning Polar Party. Two days later we were all very surprised to see Demetri return with one of the dog sledges, accompanied by Mick Crean. Mick was one of the last supporting party who turned back when Captain Scott was one hundred and forty five miles from the Pole. We gathered that Lieutenant Evans, Lashley and Crean had had a very tough time on the return journey. At the last moment Captain Scott decided to take four others with him to the Pole, this left them a man short for the return journey.

Indexed Testament presented to Francis Davies on his leaving New Zealand Nov 1910 by A. R. Falconer (Seamens' Missionary).

Lieutenant Evans had been taken ill soon after leaving the foot of the Beardmore Glacier and at times had been dragged on the sledge, in his sleeping bag, by his two companions. When within thirty miles of Hut Point the party was too weak to sledge further, so Crean volunteered to try and get help. He left Lashley to look after Evans, and with only a ski stick walked to Hut Point in eighteen hours without a break, arriving in a very exhausted condition. In spite of months of hard sledging he did the journey without food and at the risk of falling into crevasses or being caught in a blizzard, either of which would most likely have cost him his life. Fortunately Doctor Atkinson and Demetri had just arrived at Hut Point on their way South. Doctor Atkinson with one of the dog teams left immediately in a blizzard to bring in Evans and Lashley.

After they had been brought in, Doctor Atkinson sent Demetri to Cape Evans with a message for Mr Wright and Cherry Gerrard, to come to Hut Point as one of them would have to take charge of the dogs, to meet the Polar Party.

Lieutenant Evans being so ill, Doctor Atkinson could not leave him. The dogs were too tired for the return journey, so Wright and Cherry Gerrard went on foot, taking a sledge and sleeping bags, also onions and fruit for the sick man. I went with them to offer my services, as I know they had had a very heavy sledging season.

Terra Nova

Chapter XIV. Horse Racing at Christchurch

Hut Point is nearly fourteen geographical miles from Cape Evans. I was wearing ski boots without ice nails and found it very hard going over the sea ice. Before we reached Hut Point I thought my hip joints would seize up. I felt like a jointed wooden doll. About half a mile from Hut Point I fell out for a rest, whilst the others went on and came in later at my own pace, I was tired.

After we arrived at Hut Point a blizzard came on and lasted for several days and the fast ice broke out beyond Cape Evans, cutting off communications. We saw the ship in the distance on one or two occasions but she disappeared again in the drift. We were forced to wait there until we could be picked up by her.

As soon as the weather cleared Cherry Gerrard and Demetri left with the dogs to meet the Polar Party. There was not a great variety of stores in the hut but a fair supply of self raising flour, lump sugar and a few sultanas, these were eked out with seals and penguins which were killed as required. There was a very primitive stove built of stones with two iron bars across, which burnt seal blubber very well. The blubber, cut into six inch squares was laid across the bars, then a piece of paper was lit under it and the oil dripping steadily from it kept the fire going. Bill Lashley installed himself as cook and served up some tasty dishes, usually seals liver and onions followed by one of his 'specials', a chippattie, made from self raising flour and sultanas and cooked in a biscuit tin lid. The fireplace filled the hut with acrid smoke, and grease from the blubber ran on to the floor mixing with hairs from the reindeer sleeping bags, which was trodden everywhere. Everybody and everything was covered in soot.

We had to take turns for breakfast as we only had the biscuit tin lid for cooking. One morning, just as the last man was about to get his liver and onions by some means it was upset on the floor. He looked around in silence, then slowly saved what he could of the liver, scraped off the grease and hair and got on with it. In our tea we had ten lumps of sugar, Barrier rations we called it. The hut was very dismal, and in this atmosphere Lieutenant Evans lay in his sleeping bag on the floor. Yet in spite of being so ill he was always cheerful and ready with a joke, I being the only one of the party who had been in touch with the outside world for the past sixteen months, I spent a lot of time telling him all the news.

On 26 February the sea ice had broken out to within half a mile of Hut Point and the ship arrived and picked us up. The ship had had a very bad time during the fortnight she had been away. On two or three occasions she had arrived unsuccessfully to pick up Campbell and his party and had searched the coast for the western Geological party, eventually picking up the latter south of Butter Point, on 15 February. The ship, earlier in the day, had laid a depot at Butter Point and had found a note from the Geological party which had been left there only an hour or two before.

The Captain was very angry with me when I got on board and told me never to leave the ship again without his express permission. As a matter of fact the ship had gone off and left me building the stables and annexe at Cape Evans. Later in the evening he sent for me and apologised for losing his temper. It appeared the ship really had had a bad time. On one occasion she had dropped anchor and it had fouled a rock on the bottom. In trying to up anchor they had smashed the windless and had to get it up by hand with tackles, luff upon luff, as it is called which took twelve hours. When they did get the anchor up half the stock was missing.

The ship returned to Cape Evans and completed the landing of stores before making another attempt to pick up Campbell's party. It was an abnormally bad season, the sea had already begun to refreeze, pancake ice being nearly five inches thick, very viscous and full of diatoms. The ship could only make two knots through the sticky ice with the engines going at full speed. A week earlier, when she had tried to reach the party, she was frozen in the pack for forty hours and it looked as though she would have to winter there, but by rolling ship and putting the engines full ahead and full astern the whole time she eventually broke clear. The ice increased in thickness rapidly, the wind driving the pancakes one on top of the other, and in next to no time there was ice two feet thick around the ship, she was very lucky to get out.

On 2 March the Skipper had everyone aft and explained the position. The ship could do no more, she was short of coal, and on Captain Scott's orders she was not to remain in McMurdo Sound later than 4 March. The party, he said, would have to wait until the sea ice got thick enough for them to travel to Cape Evans. They had one month's provisions and would have to live on seal and penguin, of which there would probably be plenty for another month. They should reach Cape Evans, two hundred miles away in about two months.

The ship had returned to Cape Evans to report the position to Doctor Atkinson, who was then in charge of Winter Quarters. The doctor and Patsy Keohane embarked for Hut Point, where they were landed with stores and coal for the hut. We then proceeded to Glacier Tongue to ice ship, before returning to New Zealand. Leaving Glacier Tongue, we called at Cape Evans to embark those who had completed their period of service and were going home. Amongst them was Anton the Russian boy, who had been engaged to assist with the ponies. He wanted to leave badly. The winter had had a very depressing effect on him, possibly due to not being able to speak the language. He would say, 'Anton stay, Anton,' the remainder of the explanation he would make with his hands, going through the motion of putting a rope around his neck and pulling it over a beam. He had a Russian string instrument which he used to play for hours, a balalaika similar to a mandolin. Sometimes he played, and with tears in his eyes would tell us 'when Russian man play, Russian man kli (cry). Policeman come, Siberia.'

Lieutenant Evans was also on board, still very weak and unable to get out of his bunk. He was entirely dependent on the officers for medical treatment, there being no doctor on board. He was made as comfortable as possible but must have had a terrible time on the voyage. I made a sort of bed table for him that fitted securely across the bunk, with holes in one side to fit the various plates, cups etc, that he used for meals. It could be reversed and used as a writing desk. He, however, did not let things get him down, he was always cheery.

The ship had great difficulty in getting clear of McMurdo Sound. The pancake ice was eight inches thick in places, with few open leads, and clung to the ship like glue. We spent hours running from side to side of the ship to help her through the ice. Once clear of the sound and out in the Ross Sea the ice conditions were not so bad we still had gales and heavy snowstorms. It was either an abnormally bad season or the previous year had been abnormally fine. The pack-ice clung to the coast of Victoria Land the whole of the season.

As usual the ship was very light and we had even fewer opportunity of ballasting with rock than last year, such as was got on board was obtained only by tremendous effort. Often the crew had to pull sledges miles and miles over the sea ice for only a few hundredweight. I don't think we had ten tons altogether. It served another purpose in New Zealand, we gave it as souvenirs to the many thousands who visited the ship. As soon as we were clear of the land we doused fires to conserve the few tons of coal we still had for steaming into harbour. We were now entirely dependent on the sails. We had gale upon gale and the seamen had a very rotten time, particularly aloft, laying out on the yards furling the half frozen canvas.

In one particularly bad gale the ship broached to during the night in pitch darkness, and for twenty minutes her lower yards dipped in the heavy seas. The deck filled with water, well over the ice rail. It smashed the sky light over our mess and the water poured down on to the table, cascading into the bunks. The seas unhooked the for'd fall of the starboard whale boat and she hung by the after fall, banging against the ship's side, just abreast of Lieutenant Evans bunk with only the thin topsides of the poop between.

From the day we sailed I always kept two axes on the bridge for cutting away the rigging should we be unfortunate enough to get dismasted. One of the seamen dashed on to the bridge, seized an axe and cut the after fall of the boat which soon disappeared to leeward, fortunately before doing any great damage.

It appeared that when the helmsman was relieved he did not make his relief understand, probably due to the howling of the gale he could not hear, that he had taken a turn and half hitch with the lanyard that was used to hold the wheel in bad weather. Anyhow, when the helmsman tried to put the wheel over, the ship had broached to before he realised that the wheel was lashed.

I was sorry the whale boat was lost. She was a lovely model, built of pine for lightness on very fine lines and fitted out exactly as the whale boats were in the days of 'Moby Dick' and 'Frank Bullen'. We had three of them all told, they had a tank amidships where the whale line was coiled and a small decked-in foc'le on which were the two bollards for checking the line when fast to the whale. Luckily we were under lower top sails only when she broached-to. However, we got her before the wind again, the main damage being the loss of the boat and the officer's roundhouse (W.C.) had been washed inboard and lay against the saloon. The bridge – which was little more than a narrow platform across the ship, with a small match boarding shelter for the charts and navigation books was wobbling from side to side as the ship rolled. As soon as it was light enough I braced the roundhouse back into its place and fitted a shore and screw stay to the bridge, these remained to the end of the expedition.

The Captain and I struggled into the fore hold to see if the ballast had shifted. In the dim light I saw something moving with long ears, I knew I wasn't seeing things for we had been over three months at sea. It was one of the eight rabbits I had taken South and was the sole survivor. It was living on a few ends of match boarding. Whilst I was ashore one of the firemen shifted the rabbits into the 'tween deck to protect them from the weather and later the hutch had rolled into the hold and the rabbits escaped. What happened to the others I never found out. A few days before we reached New Zealand somebody caught it and brought it on deck for a feed it was literally skin and bone.

During this gale the cat was missed and never seen again. The crew spent hours of their spare time searching the ship, high and low, but found no trace of him. Strangely enough, it was about the time, almost to the day, that Captain Oates perished on the Barrier.

He had grown into a fine cat and was a great pet fore and aft. Like all the other members of the crew he had his run ashore in port, but never missed his ship, he was always on board on sailing day. He had many female admirers among the cats in the dockside warehouses of Lyttleton, by whom, no doubt, he was mourned when his ship returned without him. Like the rest of us he seemed to have no regret when sailing day hove round.

When we had the dogs on board he still roamed the deck, keeping a watchful eye on them, he seemed able to gauge the lengths of their chains to a nicety and was not unduly nervous of them. Only once did he make a slight miscalculation and that was the time he jumped overboard instead of into the rigging as he intended. A few days after the cat had been lost Billy came to me and said "Chippy I'm thinking of getting a memorial card printed in memory of our cat, what do you think of the idea?" I said I thought it quite good.

He then produced a rough draft of what he thought would be suitable. It gave particulars of his place of birth and manner of his death and stated he was the ship's mascot, but what tickled me was the verse at the end to the effect that a place was vacant at our table that never could be filled. I asked Bill if he had mentioned the matter to the Captain, he said he hadn't. I told him I thought the skipper would be upset if he wasn't asked to associate himself with the business. I said 'Why don't you go right aft and see him now. He's on the bridge keeping the first dog watch?'

As he wended his way to the bridge I gave the lads under the foc'le the tip of what was in the wind. In a few minutes Bill returned. I asked him what the Skipper thought about it. He said he was very pleased and would get some cards printed in New Zealand.

When the ship reached the milder weather of the cloudy belt it commenced to rain, so I decided to do a bit of washing, getting ready for the beach. Being in charge of the fresh water I could not set a bad example by using any myself, though I had more than suspicion that one or two of the boys flogged the galley pump quietly during the night watches, so I placed a two gallon watercan under the scupper from the poop which ran on to the upper deck. The poop was in a filthy state with coal dust, dogs' muck, blubber and chewed shrimps from the intestines of the seals that had been examined there, for as yet there had been no opportunity of scrubbing deck, but any old water that would lather soap was good enough to start the dirt, we towed our dhobeying over the side on the end of a rope to rinse it.

At dinner time I took the can of water along to the galley and put it on the range to get hot, intending to get on with the washing after tea. In the galley was a cheap and nasty pump which often lost its suction through the rolling of the ship and often required my attention. After tea I went along to the galley to get my can of water. It had disappeared. I said to Walt, the Chef. 'Do you know what has happened to my can of water?' 'Yes, made tea with it. The pump was dry and there was no water in the kettles when I came up at seven bells (15:30)' this was after he had had his afternoon shut eye.

Later the captain sent for me and asked me what was the matter with the fresh water. I said, 'There's nothing the matter with the freshwater, the tanks are in good condition.' I knew they were for I always made a point of cleaning them out and coating them with cement myself. He told me the seamen had been aft to complain and had brought a mug of tea for him to see. He told them there was nothing wrong with it, they had had the same in the saloon. The seamen held the mug out saying 'Smell it, Sir, it stinks.'

I kept mum of course, and said I could not understand what had happened, if I had spilt the beans then everybody would have been on the sick list. Many moons afterwards when we were camped in the bush during a survey of the French Pass, I told him the story of the 'Dhobeying Water'

One evening I had a little pain in my 'locker' and thought a drop out of the bottle would be the right medicine for it, so I encouraged that little pain. I thought if I forgot it would come to nothing. When I kidded myself sufficiently that I really had a pain, I went aft to see the Skipper for a dose of medicine, hoping of course, he would prescribe a tot of rum. He was very sympathetic and told me he would bring something for'd for me later. He did, a seven pound tin of white peppermint lozenges and a few magazines. It was no longer worth while nursing that pain, I forgot it.

As we approached New Zealand we had bright sunny weather with a calm sea. We passed a large school of Cachalot whales (sperm) swimming lazily with their huge heads just above water, looking like large rugby footballs. Sundays, if the weather was at all reasonable, we generally tried to spruce ourselves up a bit and tidy the mess. It is surprising what can be done with a pint mug of hot water and a little care and ingenuity. First we used it for shaving, after that we washed our faces with the shaving brush, then we poured the water, a little at a time, on the mess stool and scrubbed it with a little soap, with the remainder if any, we wiped the oiled cloth on the table and the linoleum on the deck.

We dropped anchor in Akaroa Bay on 1 April 1912. We were all very pleased to get a sight of the green fields and trees again. Lieutenant Pennell landed with Captain Scott's story for the papers, otherwise there was no communication with the shore. Almost as soon as the anchor was on the bottom we had fishing lines over the side and soon had a good catch, a welcome change to our diet. Lieutenant Pennell returned with the disturbing news that Amundsen had reached the Pole and had returned in the *Fram* to Tasmania. He reported he had seen no trace of Scott. We put to sea again as soon as the Skipper returned and entered Lyttleton harbour the following day.

Lieutenant Evans had made a good recovery and was by this time up and about and later sailed for England on business in connection with the expedition. Several other members bade adieu to the expedition including Simpson, Meares, Ponting, Clissold and Anton.

We were given a couple of weeks leave which we spent at the 'bach', the finances would not run to a holiday at a hotel. I had not yet worked off the dead horse of the previous year, we were more or less living on borrowed time. Generally speaking, we lived a very quiet existence during this period of leave, finding pleasure in the simple things of life, we had to. There was, however, one pretty hectic weekend. We mustered six in the party, two officers from a warship at Lyttleton, two of the rowing club boys, Billy Williams and myself. We started to pub crawl early on the Saturday evening and by closing time, 10 o'clock, we were 'bungs up and bilges free', as we say at sea, we took several bottles back to the bach with us and held a singsong.

During the time we had been away a bungalow had been built on the section next to ours, and the 'convenience' right up close to the 'bach', so close in fact, that there was only a chestnut fence between it and the kitchen. In the early hours of the morning one of the party suggested it was an insult to build the 'offence' so close, everybody agreed. We decided to remove it forthwith. We climbed on to the roof of the 'bach', lay on our backs and with a one, two, three heave and bust her, we pushed the sentry box, a wooden one, over on its side with our feet, leaving the 'Thunder Box' standing. When we turned in, two of the party had to sleep on the floor but we were very short of bedclothes and all we could find to cover them was a roll of wallpaper and a boat's sail. The chap under the wallpaper twisted and turned all night and the crackling of the paper kept waking the others up. In the morning, whilst strolling alongside the river, we met our next door neighbour. He said, 'there was a noisy lot of - in that 'bach' last night', indicating ours. 'Did you hear them,' why of course not.

We had invited two ladies to tea on the Sunday afternoon, but we didn't expect them to turn up. However, they did turn up, and we were quite unprepared for them. We held a meeting of ways and means, we had to give them tea.

Whilst a couple of the fellows took the girls for a walk by the river the remainder prepared the tea. I remember we were very short of table cloths, as a matter of fact I don't think we ever had any, but we managed to get over this slight difficulty by using the cleanest sheet we could find in the bunks. It was a bit grey and crumpled, I folded it neatly and put it through the sailors mangle, wrapped it around the rolling pin and then rolled it with the pastry board, putting my weight on it. It looked as though it had just come from the laundry. The tea party was a great success.

A day or so before we were due back off leave there was to be a race meeting at Christchurch. We made up our minds to attend if we could get a sub from the Skipper, so we packed our traps and returned to the ship. We got the sub without any awkward questions being asked and off we went to the races. We knew we should be about ten months of the next twelve at sea, and this would give us a chance to work back our overdraft.

It was a glorious day. Christchurch has one of the most picturesque courses I have ever been on. It was the first time I had seen the Tote operating. Our luck was out until almost the last race. In this race there was a horse running called 'Bon Ton'. I was wearing a new pair of boots and the trade mark on the tag was 'Bon Ton', I thought this a very good tip and put on my modest ten bob, the lowest stake accepted by the Tote. It came up all right and paid out odds of about six to one.

A day or two later the skipper wanted some money, and to save himself the trouble of going to the bank he asked me if I had any of the 'sub' I could let him have, I said I was sorry but was actually coming to him for another sub. He asked me if I was short of money, I told him I was absolutely broke. The Skipper did not deal with our accounts, that was done by the ship's agents ashore, he said I was to draw what I required within reason in future. So that was that.

Whilst at Lyttleton the Captain asked me if I had ever been to the hot springs at Hanmer, some ninety miles from Christchurch. Many of the crew had been invited for a trip there before the ship sailed on her first voyage South, but I was too busy to go. I replied that I had not had an opportunity. He said he was going there for a few days holiday and would be pleased if I would go with him as his guest. I said I would be delighted. He was an ideal host and splendid conversationalist. He talked of almost everything under the sun- the geological formation of the land as we drove along in the coach, the birds, animals, trees, flowers, the stars and one thousand and one subjects.

It was a delightful drive by coach for the latter part of the journey. Once we stopped at a small wayside inn to water the horse, I had a nice cool glass of beer. The country through which we passed was very hilly and almost untouched by the hand of man, a small farm here and there being the only civilization. We arrived at Hanmer about lunch time and put up at one of the temperance hotels. Almost the first thing he did was to order in a case of beer for my special benefit, he rarely drank himself, just a very little to keep me company. Each morning before breakfast we bathed in the thermal sulphur baths, the rest of the day we tramped the countryside, doing twenty to thirty miles. One day we were caught in a heavy thunderstorm in very open country and got wet to the skin.

I remember that day very clearly, particularly the conversation during the evening. He talked to me about his family and finished up by saying, 'So perhaps one fine morning you will open your paper and see that I have put my ship ashore at Brighton or somewhere, it probably won't be my fault but just fate.' Those words were in a way almost prophetic for in just four years later he went down with his ship HMS *Queen Mary* at the battle of Jutland, where he was serving as her navigating Commander. He was an officer of high ideals, absolutely selfless, who died as he had lived, serving his country in the Royal Navy. A great loss to his country and the service in which he was so proud to serve. My only son was christened Pennell after him, he met a similar fate whilst serving as navigating officer in one of the Navy's little ships in the last World War. Often I wonder if names have any influence on destiny.

Lieutenant Harry Pennell RN

Peter Pennell Davies RN son of Francis Davies

Chapter XV. Hydrographic Survey of Admiralty Bay

During the southern winter of 1912 we did a hydrographic survey of Admiralty Bay in the vicinity of French Pass between D'urville Island and South Island, New Zealand, for the New Zealand government. It was a pretty lonely spot between Wellington and Nelson, and was a haunt of the famous Pelorus Jack, a sort of black fish or dolphin, said to pilot ships through the French Pass. I believe it is the only single fish in the world protected by a special Act of Parliament, and figured largely in Maori folklore before the coming of the white man.

The story goes that one day a passenger shot at him from a ship called *Penguin*, which used to sail regularly through the French Pass. Pelorus Jack never piloted this ship again and later she was wrecked. At times it would disappear for quite long periods and then suddenly turn up again. We never saw it during the survey and I don't think it has been heard of since.

Billy Burton and I were sort of spare parts for the survey and formed the advanced party to see everything in order at the boarding house, Elmslie Bay, where all were to be accommodated, while the remainder of the party collected the two motor boats from Wellington. Not having much to do on the first afternoon after our arrival, we walked along the cliff to the lighthouse at the pass to make our number with the keeper and let him know, more or less, what we intended doing. As we wended our way along the narrow path on the edge of the cliff the deep blue water of the bay sparkled in the sunlight below us.

I said to Billy 'this is the place for blue cod', blue cod is one of the prime fish of New Zealand.

'How d' you know?' he asked.
'Well', I said 'See how blue the water is.'
'Does that make any difference?' he asked.
'Of course it does,' I replied, 'That's where they get their blue colour from.'

Billy didn't seem too sure about this information, but whatever he thought he did not let on.

After tea that evening I missed him, when next I saw him he was in a small boat anchored some distance off the shore, with a line over the side, fishing. It was now raining heavily and he was wearing oilskins and sou'westers. I shouted to him asking if he was getting any bites. He was a little deaf and after unbuttoning his oilskins and fumbling around he pulled out his watch and said, 'Ten minutes to six.' He was very proud of his watch, it was of the large turnip variety. Often I would see him mornings watching the sun rise and looking at his watch and would remark, 'Sun appears to be a bit late today, Bill.'

The accommodation at the boarding house was very comfortable if not elaborate. The beds were similar to those used by the bush whackers in their camps, just a frame and legs of poles out from the bush with sacking nailed across. We had plenty of good food but few fresh vegetables. This may seem strange living as we were in the heart of the country but all the land was given up entirely to sheep. Such fresh vegetables as we did get were brought by the steamer from Wellington, there was always plenty of fresh fish and mutton.

It was used mostly by the bush whackers who were clearing the bush in the vicinity over a widely scattered area. Here they waited to get the mail steamer that ran between Wellington and Nelson, which stopped in the bay to pick up odd passengers and small quantities of fish twice weekly.

There were only four habitations in the settlement, a bungalow occupied by the owner of the sheep station, a small store that supplied the needs of the bush whackers, a fisherman's hut and a boarding house. The sheep station was about two thousand acres, formerly bush land, which had been cleared and then carrying one sheep to the acre.

D'urville Island was still thickly covered with bush which the bush whackers had started to clear. Many of the bush whackers were 'runmen' from merchant ships. They cut the bush for thirty shillings an acre, leaving anything over four inches in diameter, for which there was a special rate according to size. The bush for the most part was of cedar, not the cedar used commercially, a soft wood up to about four inches in size and grew very thickly. When it was cut it was fired and then seeded with grass. In about six months the land was fit for grazing sheep.

The bush whackers lived a hard but simple life with very little variety in the way of food. They had to take stocks of food to last them several weeks. Apart from flour there was little more than large tins of cheap jam, occasionally they managed to get fresh mutton. They made their own bread over the camp fire, they called it damper, a kind of large pancake. Working long hours for about three months they earned a fairly big cheque, eighty or ninety pounds. Then there would be a break for a holiday in the nearest town to 'blow' the money.

Their usual practice was to hand their cheque to the hotel proprietor and tell him to let them know when it was all gone. When broke they returned to the boarding house at French Pass on their way back to the bush, obtaining credit for their digs and having to be grub steaked by the little store until they earned another cheque. When once again on their way to the bright lights they settled all their old scores in full.

We did the surveying from two large motor boats, the larger being lent by the Wellington Harbour Board, a very suitable craft for the job with respect to her accommodation, but the engine gave a lot of trouble, the other was a converted sailing boat, not quite so suitable.

We had a pretty strenuous time at the beginning of the survey putting up marks on the tops of the hills for taking angles to fix the positions of the soundings. The hills were very steep, between twelve and fifteen hundred feet high. Lieutenant Pennell and myself did most of the work on the D'urville side of the bay and Lieutenant Rennick and Bruce on the other side. Often we had to struggle to the top of one of these hills with a bucket of whitewash and needed careful 'navigation' to get it to the top without loosing it. At the top of the hill we built a stone cairn, whitewashed it so that it would show up well, and put up a flag.

Sometimes we took the theodolite and this is where I came off best for I was never trusted with the instrument itself, I carried the legs. Pennell would say if I fall with it, it will be an accident but if you fall with it, it will be carelessness. One of my jobs was to take a field board and wander along the coast searching for surveyors pegs used when the land was surveyed. I was very lucky finding these for many of them had become overgrown. When I found one I took its number, marked it on the field board and put up a distinguishing flag. These were used mostly for fixing the positions of the inshore soundings.

We worked long hours in the boat, Pennell was a glutton for work, I never saw him with time on his hands. On one occasion we had to return somewhat earlier than usual, even then the other boat had arrived alongside the pier just before us. Rennick called out 'Anything wrong, Pennell? Have you run out of flags or something?' Pulling his leg of course.

After tea of an evening there wasn't much to do bar listen to the same old records played on the gramophone night after night, wireless sets had not been dreamt of then. There was an occasional break when we received invitations to dances at one of the wool sheds, in some out of the way place, but this usually meant losing two days work by the time we got to the dance and back.

I usually gave Pennell a hand with the chart, I set the angles on the station pointers for him and he plotted the soundings. About nine thirty he would tell me to pack up, but he still carried on until about midnight. After that he read the English papers until the early hours of the morning.

There was good fishing in the bay and we had lines down at every opportunity. The water was so clear the fish could be seen swimming around, it was like looking into an aquarium. You could almost select the fish you wanted. Lunch times we landed on the beach with a good supply of filleted fish which we cooked over a driftwood fire. There were millions of shell fish in the sands on the beaches. These we cooked in the hot ashes, in their shells.

On Sundays we had a busman's holiday, we went fishing with the boats. When we had good catches we landed them at one of the bush whackers camps. We smoked them on the ubiquitous kerosene tins, flattened out, in the wide chimney of their shack. Sometimes we made up a party from the boarding house for a wild pig hunt, the pork made a welcome change in our diet from the everlasting mutton. If we came across a litter, the males were castrated and then set free. They would come another time.

A travelling parson visited the settlement about once a month and held an evening service in the dining room of the boarding house. The settlers from miles around came in their small motor boats, their only means of transport, there were no roads through the bush.

The settlers who had taken up land lived in small shacks made from poles out from the bush, sometimes with their families. Most of them were not too well off, but by dint of hard work, clearing the bush from the land for sheep and a little fishing, they carved out homes for themselves. The government looked after the education of the children and appointed a teacher no matter how few the children. She lived and taught at one of the sheep stations, her pupils coming for lessons by motor boat.

Towards the end of the survey the work lay some distance from the boarding house. Lieutenant Pennell, a seamen and myself camped in the bush at nights in order to save time getting to and from the job mornings and evenings, for there was quite a lot of the work close inshore that could only be done from a small boat. This we did after the boat left in the afternoons until she arrived in the mornings. When darkness came on we landed and cooked the evening meal, we usually looked out for an abandoned bush whackers shack, as these were fitted with rough bunks and large fireplaces for burning wood.

The evenings were long, for it was winter in the Southern hemisphere, and it was not possible to wander from the shack after dark, but they passed pleasantly enough. The seamen and I did the cooking whilst Pennell wrote up his logs by the light of a hurricane lamp. Fresh food came daily by the motor boat and we always set ourselves out to cook a good meal that usually consisted of tinned soup, fried fish, fried mutton chops with chips and a milk pudding to top off with. By the time we had dished up and done the rest of the chores it was nine o'clock , and the rest of the evening we spent around the cheery fire, spinning yarns.

One wild, blustering night we were rather disturbed when we heard a sort of maniacal laugh just outside the shack that went on for hours. We could see nothing but thought it must be one of the bush-whackers gone 'crackers'. We found out afterwards it was the call of a small species of penguin that frequented the island.

We had the choice of a salt or fresh water bathe in the morning, but we were not allowed to indulge in these luxuries for long as there were millions of sand flies. My face and arms became swollen out of shape from their bites, inspite of smearing myself with paraffin that got into my eyes with the sweat, causing them to become very inflamed.

When we had breakfast, which was prepared overnight to save time. We took to the small boat and carried on with the inshore soundings, the seaman or myself pulling the boat or taking the soundings, and the Skipper fixing the position by sextant. When the motor boat arrived we transferred to her and carried on with the deeper soundings until it was time for her to leave again in the evening.

There was a lot of hard tramping, traversing the coast line, noting the nature of the beaches etc., and this was not made easier by the fact that we always wore sea boots. One afternoon when we had already done several miles of particularly rough going, climbing over rocky points and through patches of scrub with a fair load of instruments and paraphernalia, I was thankful to hear the Skipper call the boat, which followed us along , into the beach. We had packed all the gear into the boat and I had one leg over the gunwale getting in, when he said, 'Oh, we are not getting in the boat,' so onward we tramped for another couple of miles of tough going. I thanked heaven the next day was Sunday.

We had the misfortune to lose one of the crew whilst surveying. He was drowned just off the end of the small pier and nobody knows to this day how it happened. It occurred during a weekend. He didn't show up on the Sunday but we were not unduly anxious as we thought he had gone off in one of the small motor boats to spend the day with one of the settlers. However, when he didn't turn up on Monday morning we made a search and found him at the end of the pier. We could see him through the clear water lying on the bottom. A doctor came from Nelson to hold a post mortem and inquest. Some of us had the unpleasant job of assisting. We dug a grave and buried him on a spur overlooking the bay. Later a headstone was erected and the little plot fenced off. It cast quite a gloom over our little party.

We always had trouble with the engine of the big motor boat. It was a high speed engine and the vibration was terrific. One day when we were running a line of sounding, stopping every few minutes to sound, I rang the telegraph to stop. As nothing happened I looked down the engine room hatch. The engineer looked up at me with a dazed and far away look in his eyes I saw he had been gassed and jumped down to stop the engine. We got him on deck as soon as possible and in a few minutes he was all right again. We discovered that one of the joints of the exhaust pipe had blown, filling the engine room with gas. It was fortunate the telegraph was rung or we should have had another casualty. A similar incident happened to another of the engineers shortly after, but on this occasion it was the exhaust pipe that had become fractured by the vibration. I got a holiday out of the latter incident. I took the fractured pipe to Wellington to get it welded.

Sheep shearing took place at Elmslie Bay while we were there. The sheep were all rounded up and shorn in the wool shed with machine clippers driven by a petrol engine. The shearers are experts and followed the job from place to place and can shear as many as two hundred sheep in an eight hour day.

The sheep are rounded up three times a year for crutching, which is clearing the wool away from the teats before lambing, cutting the tails and ear marking the lambs and for shearing.

One public holiday a wood chopping competition took place at one of the larger settlements some distance away, to which most of us went. The bush-whackers are naturally very keen and spend days grinding their axes to a very fine taper for the event.

The logs for chopping are fastened with iron spikes to other logs, fixed on end in the ground. One of our engineers, thinking he was being helpful, started to remove some of the blocks, after one of the chopping events, with one of the precious axes and had the misfortune to hit one of the iron spikes, making a great gap in the cutting edge. The owner was very annoyed and we had all we could do to prevent a free fight.

Francis (Chippy) Davies

Thomas Clissold ship's cook at his stove.

Mortimer McCarthy at the wheel

Photograph of officers and crew aboard Terra Nova departing New Zealand, sent as a parting gift to friends and family at home with accompanying letter.

Chapter XVI. Last Voyage South

After the survey, we returned to the ship to refit her in readiness for the last voyage south. With my seamen mates I stripped every sheave and block aloft, and the running gear was renewed as found necessary. From start to finish of the expedition never a spar or rope carried away in spite of all the gales, it had been so carefully looked after.

There was much speculation as to whether Captain Scott had reached the Pole and returned safely to Winter Quarters. We always hoped for the best and had no real reason to anticipate anything untoward happening, except that they were a little later returning than was expected. These could easily be accounted for by the abnormally bad weather we had experienced during the latter part of the summer.

In the meantime Commander Evans returned from England and took command of the ship. Whilst in England he had an audience with the King and was specially promoted to Commander in the Royal Navy.

We sailed from Lyttleton mid-December, but this time we were entirely free of deck cargo and animals, excepting a couple of cats that had strayed aboard and taken up their abode. We had been at sea about six hours when Bill Heald, who was taking a cup of soup on deck outside the galley for his stomach's sake after a pretty hectic round of farewells the previous evening, suddenly saw something wrapped in paper fall on to the ship's rail from somewhere aloft. He examined the paper very gingerly and from what he discovered he suspected there must be somebody in the lifeboat on the skids overhead, so he set off to investigate. He turned back the boat's cover and shouted 'Stowaway'.

Everybody rushed on deck to see this rare' specimen'. Two or three of the watch hauled him out and took him before the officer of the watch on the bridge. He was a disreputable looking individual, bald headed with a lock of untidy hair intended to cover the baldness hanging down one side of his head. He said he was a rabbit trapper, that was about the last job in New Zealand before becoming down and out. The Captain discussed with the officers what should be done with him. Had he been a more robust specimen of humanity no doubt he would have been welcomed as an addition to the coal trimming party, but nobody wanted him to remain, he would have been looked upon as a Jonah.

The ship was put about and headed for the nearest port, Port Chalmers to land him, but during the evening we fell in with a ship bound for Lyttleton and he was put aboard her. He had apparently come on board the previous evening, whilst the watchman's back was turned and most of the crew ashore saying their farewells before sailing. There had been some feeling during the forenoon in the foc'le about a big cake given to one of the crew by his friends ashore, which had mysteriously disappeared from the mess deck. It was found in the boat, the stowaway had taken it. On our way South we took a line of soundings, stopping each day for that purpose if the weather was favourable. The soundings were taken by the means of a Lucas sounding machine.

This machine had five thousand fathoms of piano wire reeled on a drum, and could sound to that depth. Attached to the end of the wire was a long tube, called a Bailey Rod, which brought up a sample of the bottom. At the bottom of the tube was a butterfly valve which allowed the sample to enter but prevented it being washed out again. Portable weights, varying from fourteen pounds to one hundredweight were attached to the tube to take it down. These were automatically slipped when the tube hit the bottom and, of course, lost.

This machine was operated by hand and it was long and wearisome business heaving in the wire after getting the sounding, and took four to six hours for a deep sounding of three or four thousand fathoms.

On this voyage Lieutenant Rennick had rigged up a motor boat engine, lent him by a friend, to supply the motive power for the sounding machine. A groove had been cut in the fly wheel of the engine to take a rope belt. It was a proper lash up, when after considerable trouble and much swinging of the fly wheel it could be humoured into starting, the belt would be found to have stretched and so on. In the long run little time was saved and it usually ended by having to heave it in by hand.

Passing through the pack-ice, the suction pipe to a small salt water pump on the poop, which was fixed to the outside of the ship, became damaged and partly stripped off the ship's side and it was necessary to take it off. To do this somebody had to go over the ship's side to take off the brackets which were coach screwed to the planking. I decided to do this when the ship was stopped for soundings. I got a bowline ready to go over the side and asked Jock one of the seamen, to tend it for me, he wouldn't hear of me going over the side and insisted on doing the job whilst I tended the bowline for him.

There was a heavy swell at the time and the ship was rolling moderately. Jock got over and I passed him the spanner, as he fumbled to get the spanner on the bolt head, the ship gave a roll and the water came up to his middle and I had to pull him up so that he could take off his sea boots which were full of water. Over he went again, the ship gave another roll and this time the water came up to his neck, still he hadn't been able to start the bolt. It looked odds on that I would have to do the job myself so I encouraged him to stick it and eventually he got the bolt out.

When I got him inboard again I said, 'You had better come along and have a tot of whisky, Jock.' I had been made a present of a case before leaving Lyttleton. I was taken aback when he said, 'No thank you, Chippy', for generally he would have given his soul for a tot.

'What's biting you?' I said.
'No, Chippy,' he said, 'I've knocked it off.'
'Go on', I said, 'Forget it, it will drive the cold out.'
'No, I'm not going to have any more and if you see me in twenty years time I shall still be a teetotaller.'

I'm afraid his good resolution did not last, for as soon as the ropes were made fast the next time in Lyttleton, Jock shaped course for the nearest pub. I saw him that night sleeping on the galley deck, in all the grease and muck dressed in a new suit he had bought that same afternoon.

Jock was a character, a real blue nose sailing ship man. He was born in the Hebrides, we could only guess how long ago, for his ship discharges conveyed nothing and had long since fallen back on the old sailor's dodge of putting his age back every time he signed articles on a new ship. He joined the ship at Lyttleton on her first voyage South and was paid off after the second voyage, more or less at his own request. He came to me one morning and asked me to see the Skipper, to pay him off. He had been the watchman the previous night and sported a lovely black eye. He told me the story that somebody had tried to force his way on board during the night and he had fought him off receiving the black eye as a reward for his bravery. I had more than a suspicion that he had been chased by a pink elephant and had fallen on one of the spokes of the wheel.

I feel quite sure that the main reason he wanted to be paid off was the prospect of the ship being laid up for a few months, we did not know then we were going to get the Admiralty Bay job. On being paid off he immediately signed articles on one of the coasters. I don't think he could ever have fitted into any job ashore.

He knew all the portents and had all the superstitions of the sea. It wasn't necessary for him to make a synoptic chart to predict the weather, he could tell from other signs, in particular the way the birds behaved. Often he would say to me, 'It's going to be a three day gale, Chippy,' or maybe it would be a seven day gale or that the gale was going to break as the birds were flying to windward, he was usually right.

When one of the lads brought a skull on board, which he had found on a desolate part of the shore in New Zealand, Jock was very perturbed and predicted some dire calamity, it was, as a matter of fact, about the time the party set off for the Pole. He came back to the ship before she was due to sail on her third voyage South in a very worried frame of mind, and asked me if I thought the Captain would sign him on again. I told him I was quite sure the Skipper would be only too pleased to have him back. However, he was not satisfied until I saw the Captain on his behalf and he was again signed on. Jock never wore boots on board if he could possibly do without them. Outside the foc'le was a ring-bolt which was right in the gangway. Every time he passed in or out of the foc'le he kicked the ring-bolt with his bare feet and every time he swore the same oath. In the night watches we would know that it was his turn at the wheel by the rattle of the ring-bolt as he went aft, it seemed to have some magnetic attraction for him.

The Scott expedition, however, was not to be his last Antarctic expedition for he was with Shackleton when his ship was crushed in the ice and lost in the Weddell Sea, and was one of the party marooned on Elephant Island. When Shackleton with five companions made the perilous journey from Elephant Island to South Georgia, in a ship's lifeboat, to bring them relief. Mick Crean and another seaman of the Scott expedition, McCarthy, were with Shackleton in the boat. Jock served with the army during the last Boer War and was captured and made a prisoner of war.

All Christmas Eve it blew like stink, we were practically under bare poles. Preparations which were being made for the Christmas festivities had to be abandoned. The poultry hung outside the galley looking very bedraggled from the spray that constantly broke over them, one of the geese had been partly plucked, there were also a couple of turkeys and two fowls. The gale still raged throughout Christmas Day, but this did not entirely dampen the spirits of the crew under the foc'le. Christmas parcels were opened and produced all sorts of good cheer, puddings, cakes, sweetmeats and a few bottles of the best. Everyone received a card and present from Mrs Scott.

In the afternoon the boys had a sing song on the mess deck. I took the wheel for a couple of hours to let the helmsman join the party, everybody was mellow and happy. The stowaway had brought a tiny wild rabbit on board with him, so small that it was kept in a wire rat trap, and in spite of the unusual conditions seemed to thrive. As a special treat somebody put it on the mess table amongst all the good things to help itself. It sat with its fore paws on a loaf happily nibbling away and did not appear to be affected in any way by the motion of the ship. The weather moderated by the evening so the ship was topped to take a sounding. The biologist also made it a quantitive plankton station, taking several hauls between four and six hundred fathoms with the silk nets.

On Boxing Day we passed several large tabular bergs, the first sighted for the voyage. We stopped again to take a sounding and plankton samples and also towed the young fish trawl.

These operations on this voyage were a daily routine, weather permitting. On 29 December we reached the pack-ice, at first fragmentary and later soft and slushy. At midnight on 31 December the old year out and the New Year in custom was carried out in the time honoured fashion at sea, sixteen bells were struck on the ship's bell by the youngest member of the ship's company. Everybody sang 'Auld Lang Syne' accompanied by the 'squeegee band,' a mandolin and a home made one stringed violin, a mouth organ and a large sugar tin for a drum. The band afterwards serenaded the officers, not without an eye to the possibility of the Skipper 'Splicing the Main Brace'.

The Christmas Day festivities, which were postponed owing to the gale were celebrated on New Year Day. We had turkey, plum pudding and plenty of beer, cigars, cigarettes, fruit etc. At midday the pack was so heavy that we had to stop and the ship was anchored to a large floe. In the afternoon ski races were held on the floe, in one of the heats I could not stop myself in time and went over the edge of the floe, dropping about three feet on to the very thin slushy ice. Fortunately I landed upright on my skis, the ice through which I was gradually sinking just held until I could be pulled out. I didn't enjoy the situation as the killers had been blowing between the floes only a few minutes before.

With the rest of the Warrant officers, I dined with the Captain and Officers in the saloon. It was a very lively party. Everyone was called upon to sing, or at least make a noise or pay a forfeit. Lieutenant Pennell who like myself was a Devonian, was called upon to sing the 'Devonshire National Anthem,' Jan Stewer and to be accompanied by 'Jan' Davies, navigator, the Skipper was slinging off a bit and dubbed me Davies Carpenter (N) the (N) standing for navigator in the Navy List and referred to my being Pennell's assistant during the Admiralty Bay Survey. After leaving the saloon we found the foc'le singsong still going full bore, so we joined in until the early hours of the morning.

We were nearly a fortnight working through the pack-ice in the Ross Sea, but everybody was kept busy, particularly the seamen who trimmed coal from the main hold to the bunkers daily, all the baskets having to be whipped up by hand.

Bill Heald always brought the gramophone along and played records from 'Maid of the Mountains,' 'Dollar Princess' or 'Belle of New York', reminiscent of the band on coal ship days in the Royal Navy. He also provided the party with peppermint Bullseyes. I don't know where he got them from but he produced an endless supply of seven pound tins. He also nailed a tin to the bulkhead in the foc'le and always saw that it was well supplied.

We had fog occasionally, which froze on the rigging, forming into icicles and fell in showers on the deck as the ship bumped her way through the ice. The ship arrived in McMurdo Sound on 18 January 1913 and was able to anchor quite close to Cape Evans, which was clear of sea ice.

Commander Evans hailed the shore party, asking if all was well. Lieutenant Campbell replied, and informed us that Captain Scott and Polar Party had reached the South Pole on 17 January 1912 and all had perished on the return journey from exposure and want. He also reported that he and his party and the other members of the expedition were all well. There was a deadly silence, broken only by the noise of the skua gulls.

Soon we heard the full story. After the ship had left in March, Doctor Atkinson and Patsy Keohane laid out depots from Hut Point for Campbell and his party who were expected to sledge to Winter Quarters as soon as the sea ice was thick enough. Cherry Gerrard and Demetri with the two dog teams, reached One Ton Depot where they were held up for a week by a blizzard and had to return to Hut Point without seeing anything of the Polar Party.

After laying out a depot at Butter Point for Campbell's party, Doctor Atkinson and Patsy Keohene sledged out to Corner Camp and depoted another week's provisions. On 30 October, 1912 a search party, led by Doctor Atkinson with Mr Cherry Gerrard and Demetri, left with two dog teams, and a day or two later a party followed with the seven mules, with supplies for three months. On 12 November, eleven miles south of One Ton Depot, the tip of Captain Scott's tent was seen above the snow. Inside it they found the bodies of Captain Scott, Doctor Wilson and Lieutenant Bowers and also records giving information concerning the other two members of the party.

First death was that of Edgar Evans, who died on 17 February at the foot of the Beardmore Glacier. His death was accelerated by concussion of the brain, sustained while travelling over rough ice on the Beardmore. Captain Oates was the next to be lost. His feet and hands had been badly frostbitten and although he struggled on until the 16 March, when the party reached 80 deg 30 of the latitude, he could go no further.

On the 17 March, his thirty second birthday, he walked out of the tent saying 'I am just going outside and may be some time'. His comrades knew that it was the act of a brave man who gave his life in order that his companions might have a chance to get through. Before leaving the tent he had spoken of his mother and his regiment.

The logs, private letters etc. were collected and the tent was lowered over the bodies. A large cairn of snow blocks was built to mark the spot and a cross erected overall. The wording was as follows:

THIS CROSS AND CAIRN ERECTED OVER THE REMAINS OF:

CAPTAIN R.F. SCOTT CV RN
DOCTOR E.A. WILSON
and
LIEUTENANT H.R. BOWERS RIM

AS A SLIGHT TOKEN TO PERPETUATE THEIR GALLANT
AND SUCCESSFUL ATTEMPT TO REACH THE POLE.
THIS THEY DID ON JANUARY 17TH, 1912,
AFTER THE NORWEGIANS HAD ALREADY DONE SO ON
DECEMBER 1ST, 1911.

ALSO TO COMMEMORATE THEIR TWO GALLANT COMRADES,
CAPTAIN L.E.G. OATES, OF THE INNISKILLING DRAGOONS,
WHO WILLINGLY WALKED TO HIS DEATH IN A BLIZZARD
ABOUT TWENTY MILES SOUTH OF THIS PLACE, TO TRY
AND SAVE HIS COMRADES BESET BY HARDSHIP,
also of
PETTY OFFICER EDGAR EVANS, WHO DIED AT THE FOOT
OF THE BEARDMORE GLACIER
THE LORD GAVE AND THE LORD TAKETH AWAY.
BLESSED BE THE NAME OF THE LORD.

The search party then travelled South to look for the body of Captain Oates but no trace of it was ever found. A cairn with cross of skis over was erected near the spot where he left his comrades, and the following record left on the cairn:

HEREABOUTS DIED A VERY GALLANT GENTLEMAN –
CAPTAIN L.E.G. OATES
INNISKILLING DRAGOONS,
WHO, ON THE RETURN FROM THE POLE IN MARCH 1912.
WILLINGLY WALKED TO HIS DEATH IN A
BLIZZARD TO TRY AND SAVE HIS COMRADES
BESET BY HARDSHIP.

I made a large wooden cross, in sections, to be erected on Observation Hill, seven hundred feet high overlooking the Great Ice Barrier. The cross took me thirty hours to complete without a break, except for meals. After I had had a sleep, a party with Doctor Atkinson in charge, set off with the cross on two sledges for Hut Point. It was very heavy going over the sea ice as there had been a tremendously heavy fall of snow. After four hours we camped for lunch.

When within a mile of Hut Point we struck rotten ice, Crean who was pulling opposite to me in the sledge went through to his waist and the rest of us put our legs through several times. It was only by keeping on the run that we were able to get clear of the bad patch, which was little more than a layer of half frozen snow. The following sledge stopped until they saw we were all right again, then made a detour around the bad patch. The journey to Hut Point took us eight hours. We soon had the primus and cooker going for pemmican and hot tea and changed our wet clothes.

A party then went to the top of Observation Hill, a mile south of Hut Point, to dig a hole in the rock to set the cross, as no time was to be lost if we were to get back to the ship before the sea ice became too rotten.

The Polar Party having reached the Pole

Meanwhile I put the finishing touches to the cross, painted it white and blacked in the lettering that had been cut into the wood. The party on top of the hill had a slice of luck and completed the hole by seven o'clock, when we were called down for supper. We turned in our sleeping bags fairly early in order to make an early start in the morning. By eleven the following morning the cross had been taken to the top of the hill which was a very stiff climb. Here I bolted all the parts together and erected it in three hours. Its inscription is as follows:

IN
MEMORIAM
CAPTAIN R.F. SCOTT RN
DOCTOR E.A. WILSON, CAPTAIN L.E.G. OATES Ins Drgs
LIEUTENANT H.R. BOWERS RIM, PETTY OFFICER E. EVANS RN
WHO DIED ON
THEIR RETURN
FROM THE POLE
MARCH 1912
TO STRIVE TO SEEK
TO FIND AND NOT
TO YIELD.

We gave three cheers for Captain Scott and his companions, photographs were taken and we returned to Hut Point and started off for the ship as soon as the sledges could be packed.

Mrs Scott with her son, Peter

Captain Robert Falcon Scott CVO RN

Robert Falcon Scott (1868 - 1912)
His life and connections to the City of Plymouth

Robert Falcon Scott was born on 6 June 1868 at the family home, Outlands, in Plymouth, where he grew up. His father managed the Hoegate Brewery on the Barbican. There are accounts of Scott learning to sail with his father and brother in Plymouth Sound in an 18 foot boat with a big lug sail. At the age of 12, Con as he was known to the family was sent to Stubbington School near Fareham to be prepared as a candidate for HMS Britannia College at Dartmouth. Having passed the entrance examinations Scott began his naval career in 1881 as a 13 year old sea cadet. At the time, the college was based aboard two old hulks moored on the River Dart and life there was harsh. In 1883 he passed out from Britannia as a midshipman.

It was as a young midshipman that he greatly impressed and was duly noted by Sir Clements Markham who was looking for young ambitious naval officers to command a British Expedition to the Antarctic, which he was beginning to plan. He placed Scott on his shortlist.

In 1891 Scott returned for three years to Devonport Naval Base to learn the new skills of torpedo warfare. This coincided with a financial crisis in his family, worsened further by the deaths of his father and brother. This resulted in the young naval officer having the full responsibility to provide income for his mother and sisters.

Scott's naval career proved successful and in 1899 he applied for and was appointed commander of the National Antarctic Expedition, *Discovery* 1901, under the leadership of Sir Clements Markham. On this expedition, 'Farthest South' was reached and much pioneering scientific work was carried out including gathering extensive data on magnetism, meteorology and marine biology.

In 1908 Scott married Kathleen Bruce, a sculptress. Their only child Peter Markham was born in 1909, he was to become Sir Peter Scott, one of the founders of the World Wildlife Fund, now known as the World Wide Fund for Nature.

Following the success of his first expedition Scott was keen to plan a future journey South. In March 1909 he took the Admiralty based appointment of naval assistant to the Second Sea Lord which placed him conveniently in London. In December he was released on half pay, to take up full command of the British Antarctic Expedition, *Terra Nova*, which set sail from Cardiff in June 1910. This expedition had the twin aims of being first to reach the South Pole and carrying out a wide programme of scientific research.

Captain Scott and the Polar Party, having reached the South Pole in January 1912 only to find they had been beaten by the Norwegian explorer Roald Amundsen, perished on their return journey. It is believed Scott died on or about 29 March 1912 on the Ross Ice Shelf, aged 43.

It is recorded that Scott returned to Outlands just before he left England to sail for the last time to the Antarctic aboard *Terra Nova*. On this visit he carved 'Scott' on a birch tree he had long before planted in the grounds. After his death, the carving was preserved and mounted, this is now on display in the new Plymouth City Museum, The Box.

The National Memorial to Scott and the Polar Party is located at Mount Wise, Devonport overlooking Plymouth Sound and was unveiled in August 1925.

May 21, 1913 — THE DAILY MIRROR — Page 7

IN MEMORIAM: CROSS ERECTED BY THE OFFICERS AND MEN OF THE EXPEDITION.

Cross erected on Observation Hill to the memory of the five heroes. The hill overlooks the winter quarters of the Discovery, the ship of Captain Scott's first expedition, and commands a clear view over the great ice barrier.—(Copyright in England. Droits de reproduction en France reservées.)

The Memorial Cross constructed by Francis Davies still stands today on top of Observation Hill.

Chapter XVII. Return to Cardiff

The ship was about ten miles distant. For the first few miles the ice was very rotten but by the time we got on to firmer ice we had a strong breeze, and being a fair wind we set sail on the sledges using the tent bamboo for a mast and the floor cloth of the tent for a sail. We made good progress, nearly two miles an hour. After about three hours the wind dropped again and we had to take down the sail. We camped and made tea, eventually reaching the ship about eleven p.m. I had had enough for one day, I was dead beat.

Whilst we were erecting the cross, stores and scientific specimens were being transferred from Winter Quarters to the ship, and by the time we returned was practically complete. The following morning Winter Quarters were secured and the ship left for Granite Harbour and Terra Nova Bay to pick up specimens cached by the Western geological party and Campbell's party, en route for Lyttleton. Doctor Nelson made me a present of a seal embryo, preserved in spirit and Demetri a very fine specimen of weathered keynite as souvenirs.

When Lieutenant Campbell realized that the ship would be unable to pick his party up, he made preparations to winter. With their ice picks they dug out an igloo in a snowdrift, the compartment being less than twelve feet square. There were not many seals or penguins on which they had to depend mainly for food, the penguins had mostly migrated to the north and the seals did not come up on the ice owing to the very bad weather, but they always kept their ice holes open so they could breathe. On rare occasions, during calm weather, old seals were seen on the ice and every effort was made to catch them.

Tiny Abbott, while endeavouring to catch one of these seals, had the misfortune to cut the fingers of his right hand to the bone, in a few minutes the blood congealed and froze inside his mit. He had chased the seal close to its hole and in trying to prevent it escaping jumped on its back and stabbed quickly with his knife. The point of the knife struck a bone in the seal and his hand slid down over the blade. When eventually it healed his fingers were stiff. Being a physical training instructor this worried him quite a lot, more so perhaps because of the abnormal conditions under which they were living.

One of the party, no name, thinking perhaps to ease his mind, said he would use his influence when they got back to get him a job as a tram conductor. Brownie treated this as a huge joke and would call out, 'Tiny, ding, ding' going through the motions of pulling the bell, there were few push bells then. He kept the joke up all the voyage home, if there happened to be a loose piece of spun yarn or suchlike hanging down anywhere he would get hold of it and shout 'Tiny, ding, ding.'

One of the seals provided a real change in their menu. When it was opened up they removed over thirty fish from its stomach in various stages of digestion which they fried in blubber. Each man had a little blubber lamp made from empty tins for light and reading, these gave off a tremendous amount of smoke and soon the igloo was inches thick in soot. They had contrived a larger stove for cooking. Nothing of the seals was wasted and the bones were saved in case it became necessary to grind them into flour if other sources of food became short.

They had no changes of clothing, hair cut or shave for nearly ten months. Naturally a lot of the conversation was mostly about food, what they would have when they got back again to civilization. There was one celebration they arranged to hold in a certain fried fish shop in Pompey. Tiny suggested that for each fish they had they would have a pint of beer. Brownie said, 'In that case I shall order whitebait.'

Seal embryo preserved in spirit and given by Dr Nelson to Davies as a souvenir.

They had very little in the way of reading matter, just two books of Charles Dickens which they took in turns to read aloud. Brownie knew lots of passages off by heart. Coming back in the ship he was cook's mate and Tiny was my mate. At most unexpected times he would suddenly emerge from the galley and quote a line or two.

They had no tobacco but managed to get a smoke from bits of seaweed picked up on the beach, shavings that had been used for packing in some of the cases, venesta three ply cases and the sledge meter I had made for them before they landed.

At the end of September they set out to sledge to Winter Quarters, Cape Evans. They found the depot of food that Doctor Atkinson had left for them at Butter Point. Here they took a day off and had a real blow out of biscuit, butter and lard. From Butter Point they attempted to make direct for Cape Evans across McMurdo Sound over the sea ice, but after going nine miles the ice became so rotten they had to turn back. They arrived at Hut Point early in November shortly after the search party had gone south to look for the Polar Party. Here they found messages from Doctor Atkinson informing them of the fate that had overtaken Captain Scott and his brave companions. It was the first news they had had of the Cape Evans party for more than twenty months.

Being earlier in the season than the two previous voyages, the weather on the return to New Zealand was moderately good. There were only two outstanding incidents, when we ran into a cul-de-sac between very high bergs during fog, and when we passed an enormous ice berg twenty one miles long, up to that time it was largest berg ever sighted.

The ship arrived at Timaru, New Zealand, on 11 February, from where a cable was sent giving the tragic news to the world. We put to sea again for twenty four hours before entering Lyttleton with our flag at half mast.

A memorial service was held in Christchurch Cathedral attended by the ship's company.

Two days after our arrival from the Antarctic I accepted an invitation to spend a weekend at Port Levy, just outside Lyttleton Heads, in a small sailing boat. Port Levy was a small undeveloped harbour. There were only a few shacks around the bay and these were mostly occupied by Maoris. The boat was very small without any cabin accommodation. We slept in sleeping bags on either side of the centre board and went ashore to wash in a stream in the mornings. There was plenty of good fishing both by line and net. We cooked our own meals over a primus stove, fried fish being the main course. There were many holiday makers in small boats, like ourselves, so a regatta was organised and all the boats took part. We won a prize given by one of the local inhabitants, bottled gooseberries.

I'm afraid I did not appreciate that weekend to the fullest for we had a head wind with a heavy sea most of the way back to Lyttleton. I spent a few hours in the dinghy battling with the seas, drenched with spray, towing the yacht. I began to think there must be something queer in my make up to take on such a job for relaxation, after months in the Antarctic.

The pipe of the little stove fitted in our mess ran along under the desk for about six feet and about six inches from it,and was a convenient place to store coats etc. One morning the engineer lit the fire and went off to his work in the engine room forgetting to see if there was anything hanging over the pipe, the chef's overcoat was hanging there. When I came into the mess for lunch the engineer was contemplating the remains of the coat which had been scorched right down the back.

This was too much for me, the situation struck me as being funny, I roared with laughter. Suddenly a voice shouted from the skylight, 'I know what you 'baskets' are laughing at, I can't see anything to laugh at etc,etc. 'He had evidently seen the state of his coat before we had and was waiting on top to hear what we would have to say when we saw it. He was very annoyed, naturally.

We left Lyttleton for Rio-de-Janeiro and Cardiff in the middle of March 1913. The weather was very bad between Lyttleton and Cape Horn. We were making a composite great circle track to shorten the passage, which took us amongst the gales and ice bergs once more. We intended to use sail only when we got into the Westerlies, but the following seas were so heavy they threatened to poop the ship and we had to keep going under sail and steam. Commander Evans and several of the officers and scientists left the ship at Lyttleton and came home by mail boat. The ship was commanded by Lieutenant Pennell for the homeward voyage.

One Sunday morning when we were approaching Rio, there was a devil of a 'chemosel' in the galley, both the steward and the chef giving vent to their feelings in a beautiful flow of nautical phraseology. I came on deck to see what it was all about. They told me the pump had lost it's suction again and in spite of recharging it many times nothing happened. I tried it and found that it was choked. I gave it a few sharp pumps and told the steward to put his jug under. In a few seconds the pump 'gave forth', there were bits of fur, a kitten's leg and pieces of decomposing flesh. Some time before one of the cats had given birth to four kittens and two of them had disappeared, we thought she had eaten them as cats sometimes do but now we knew where one of them had gone.

The doctor spoke to me about it and said, 'You can't use that water, you will have to switch to another tank.' I told him I had already done so. As a matter of fact all the other tanks were empty and there was barely enough water to last us to Rio. I think I had the worst of the show, every time I thought about it I retched.

We spent three days in Rio. It was a beautiful city with its fine buildings and lovely mosaic pavements. It is one of the finest harbours, if not the finest in the world.

As we approached at night it was a wonderful sight with its millions of twinkling lights. We could smell the perfume either of flowers or the artificial variety but nevertheless quite definite and pleasant on the night air miles out to sea. The city itself smells of roasting coffee beans.

There is a marvellous promenade with three fine roads separated by grass plots and trees, two for one way motor traffic and one for heavy traffic. This was the first time I had seen motor cars in any great number, they were stopped at a traffic control in one of the main streets, four abreast, there were hundreds of them. We visited the famous Botanical Gardens and were invited by the British Consul for a trip up Corovada, by mountain railway. We gorged ourselves on the luscious tropical fruits. The ship was coaled and watered for the last time before reaching Cardiff. It was very pleasant sailing in the steady breezes of the Trade Winds in the Tropics, after the gales of the Antarctic.

We called at one of the Western Isles to send a cable giving the approximate date of our arrival at Cardiff, only the Skipper landed. From the ship the island looked like a large market garden, entirely devoid of trees, with many windmills dotted about for pumping water. The Scilly Isles was our last port of call to complete the smartening up of the ship ready for Cardiff and to give the members of the expedition, already in England, time to reach Cardiff to meet her.

A fine reception was given us by the people of Cardiff and we were entertained by the Lord Mayor and the members of the Stock Exchange. A day or two later everybody went their separate ways. I was specially promoted for my services and returned to the Royal Navy.

I was shipmates again with only one member of the expedition, the second engineer. We served six years together in another Antarctic Scientific Research Expedition after the First World War.

LIST OF MEMBERS OF THE EXPEDITION:
CAPTAIN ROBERT FALCON SCOTT CVC RN
LIEUTENANT E.R.G.R. EVANS RN
LIEUTENANT V.L.A. CAMPBELL RN
LIEUTENANT H.L.L. PENNNELL RN
LIEUTENANT H.E. de P. RENNICK RN
LIEUTENANT W.M. BRUCE RNR
LIEUTENANT H.R. BOWERS RIM
CAPTAIN L.E.G. OATES, 6th Inniskilling Dragoons.
SURGEON G. MURRAY LEVICK RN
SURGEON E.L. ATKINSON RN
ASST. PAYMASTER F.R.H. DRAKE RN (retd)
CHIEF ENGINEER W. WILLIAMS CERA RN
SECOND ENGINEER W.A. HORTON ERA RN
THIRD ENGINEER W. LASHLEY CHIEF STOKER RN
BOATSWAIN A. CHEETHAM RNR
CARPENTER F.E.C. DAVIES, LEADING SHIPWRIGHT RN
CHIEF STEWARD W.W. ARCHER

PETTY OFFICERS:
EDGAR EVANS RN
R. FORDE RN
T. CREAN RN
T.S. WILLIAMSON RN
P. KEOHANE RN
G.P. ABBOTT RN
F. PARSONS RN
W.L. HEALD RN
J.H. MATHER RNVR
F.V. BROWNING RN
A.S. BAILEY RN

LEADING SEAMAN:
A. BALSON RN

ABLE SEAMAN:
H. DICKASON RN
J. LEESE RN
R. OLIPHANT RN
T.R. McLEOD RN
M. McCARTHY RN
W. KNOWLES RN
C. WILLIAMS RN
J. SKELTON RN
W.H. McDONALD RN
J. PATON RN

LEADING STOKER:
R. BRISSENDEN RN
E.A. McKENZIE RN
W. BURTON RN
B.J. STONE RN

FIREMAN:
A. McDONALD
T. McGILLON
C. LAMMAS

COOK:
T. CLISSOLD RN

STEWARD:
F. J. HOOPER
W.H. NEALE

GROOM:
A. OMELCHENKO

DOG DRIVER:
D. GEROF

SCIENTIFIC STAFF:

CHIEF OF THE SCIENTIFIC STAFF and ZOOLOGIST:
EDWARD ADRIAN WILSON BA MB

METEOROGIST:
G.C.SIMPSON D Sc

GEOLOGIST:
T.GRIFFITH TAYLOR BA
F.D. DEBENHAM BA
R.E. PRIESTLEY

BIOLOGIST:
E.W. NELSON
D.G. LILLIE, MA

PHYSICIST:
C.S. WRIGHT BA

PHOTOGRAPHER:
H.G. PONTING

IN CHARGE OF DOGS:
C.H. MEARES

MOTOR ENGINEER:
B.C. DAY

ASST. ZOOLOGIST:
A. CHERRY GERRARD

SUB-LIEUTENANT Norwegian NR Ski Expert:
T. GRAN

IN CHARGE OF MULES IN SHIP:
J.R. DENNISTOUN

Notes:

Chapter 2
*1. Tom Crean, from County Kerry in Ireland, is called 'Mick' throughout the book.

Chapter 5
*2. Davies was mistaken in the origin of the 'Birdie' song, it was in fact one of the songs from 'The Morning'. 'Birdie' (J.D. Morrison).

Chapter 10
*3. Built in early 1899 where 10 men led by Carsten Borchgrevink spent the first winter on Antarctica.

*4. This is at 69 deg 18' S and 158 deg 34' E.

*5. Chief Steward Walter Archer later became sole cook, when Thomas Clissold suffered concussion in a fall

Herbert Ponting

Davies far right

ROSS ISLAND AND SURROUNDING AREA
(Not to Scale)

SOUTH POLE
450 MILES SOUTH
OF
BARRIER END
OF
BEARDMORE GLACIER

BEARDMORE GLACIER
Mt. DISCOVERY Mt. MORNING
MINNA BLUFF
KOETTLITZ GLACIER
GREAT ICE BARRIER
ONE TON DEPOT
ROUTE OF POLAR PARTY
WHITE ISLAND
BLACK ISLAND
EDGE OF GREAT ICE BARRIER
Impassable Crevasses and Cliffs
HUT POINT
Glacier Tongue
Mt. EREBUS
13,350 ft.
BARNE GLACIER
CAPE EVANS
CAPE BARNE
Mt. TERROR
11,290 ft.
R O S S   I S L A N D
CAPE ROYDS
McMURDO SOUND
Open Water in Summer
Frozen in Winter
CLIFFS OF CAPE CROZIER
Penguin Rookery
EDGE OF GREAT ICE BARRIER
CAPE BIRD

ROSS SEA

APPROXIMATE DISTANCES—
CAPE BIRD to CAPE EVANS . . . . . 30 miles
CAPE EVANS to HUT POINT . . . . . 15 miles
CAPE BIRD to CAPE CROZIER . . . . 50 miles
HUT POINT to BEARDMORE GLACIER . 450 miles

Lieutenant Commander Francis Davies RNVR

Antarctic Greetings Cards

Ephemera collected and made by Francis Davies connected to his time on the Terra Nova Expedition

Invitation Cards

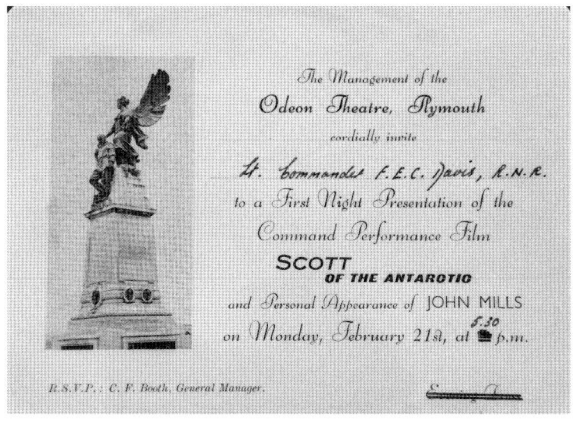

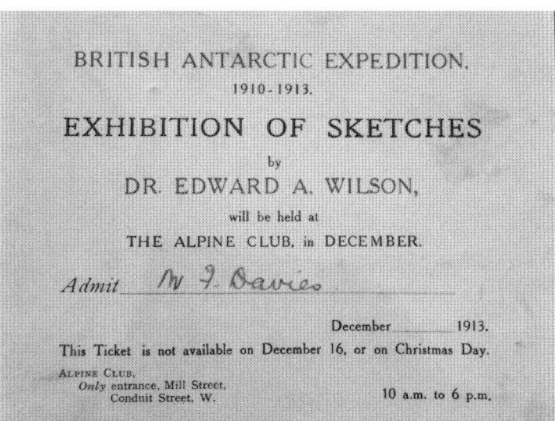

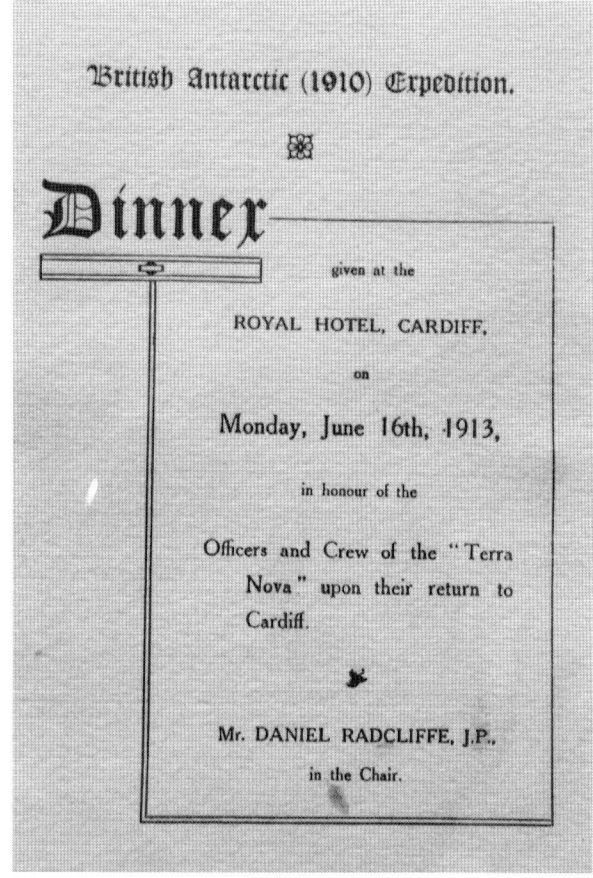

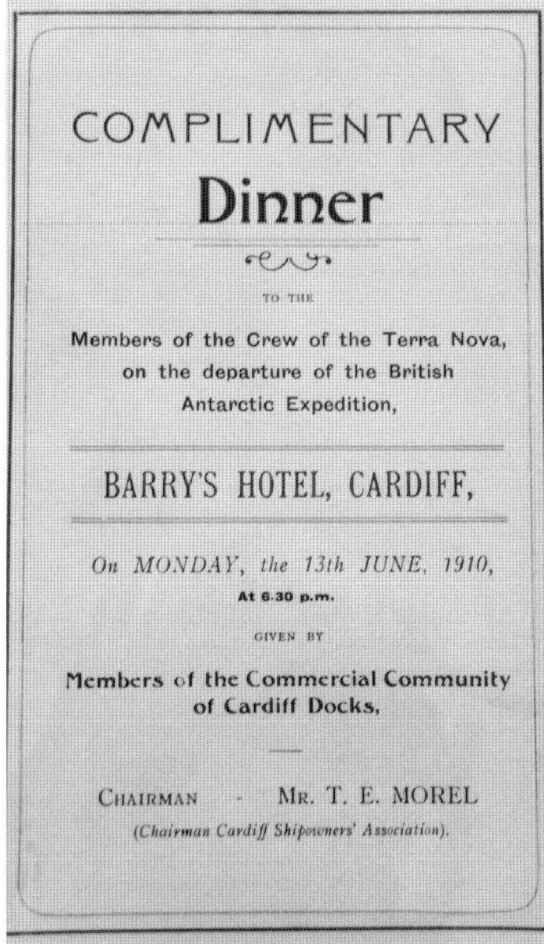

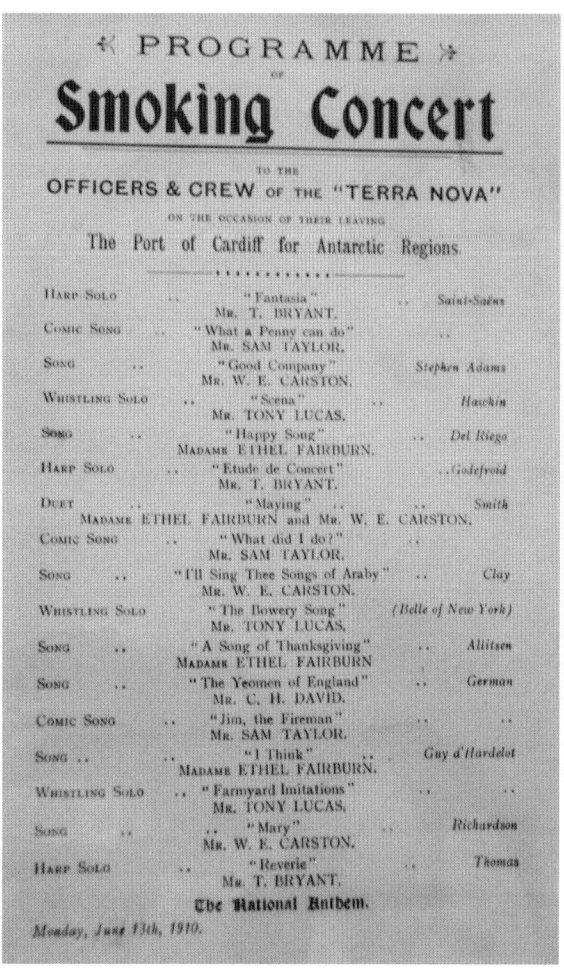

Letters

Awliscombe
Honiton

Dear Davies

I wonder if you have heard of Commander Pennell's engagement. We are all so delighted with the Lady of his choice. We do not know when the wedding will be. He comes home to-morrow for Xmas & then goes to Switzerland where Miss Hodson is

staying at present
May every good go with you through 1914.

Your sincere friend
Winifred Pennell

---

15 Queen Anne St
Cavendish Sq
W.

Dec 31st 1913

Dear Davies

Very many thanks for your letter.

So you see I am not the confirmed bachelor you used to fear I was.

I go off to Switzerland next week when the work with the ship's records is finished to take my leave there.

An engagement means being photographed & so one will be sent you when the photographer delivers them

Yrs very sincerely
Harry Pennell

Letters

Godford Park House
Awliscombe
Honiton
July 9th

Dear Mr Davies,

Knowing what an interest you take in my eldest sister, I am sending you a magazine which tells all about the work in Central Africa, in which she is taking part. This month is especially interesting to us as there is a letter in it from her & that is why I have sent a copy to you.

If by any chance you would like me to send you the magazine every month I shall be delighted to do so, as each book sold is a help to the Mission. The books cost 6d a year or 3d ½ year, & the postage is the same, so it only comes to 1/- all told for the whole year.

We have very cheery letters from my brother, he is in the Defence just now, on account of one of the officers falling ill, but expects to rejoin the Duke of Edinburgh before this month is out. He & Miss Hodson quite hope to be married this winter, in which case they will be married in Malta. I met Mr Maers yesterday he was lecturing at Exeter, & seems very well. I hear Abbot has quite recovered. I am glad.

Yours truly
Dorothy L. Pennell

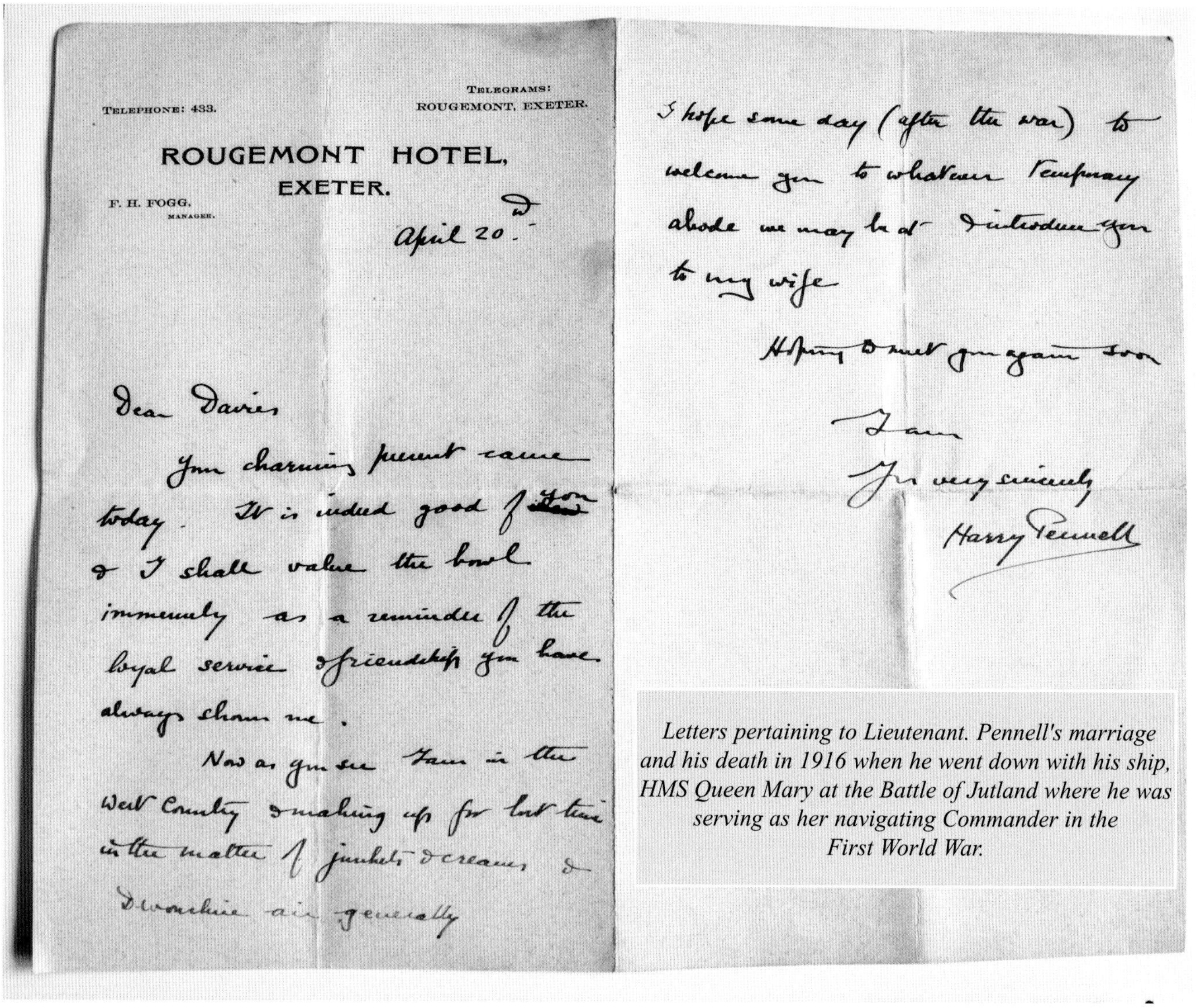

Letters

> ROUGEMONT HOTEL,
> EXETER.
> F. H. FOGG, MANAGER.
> TELEPHONE: 433.
> TELEGRAMS: ROUGEMONT, EXETER.
>
> April 20th
>
> Dear Davies
>
> Your charming present came today. It is indeed good of you & I shall value the bowl immensely as a reminder of the loyal service & friendship you have always shown me.
>
> Now as you see I am in the West Country & making up for lost time in the matter of junkets & creams & Devonshire air generally
>
> I hope some day (after the war) to welcome you to whatever temporary abode we may be at & introduce you to my wife
>
> Hoping to meet you again soon
>
> I am
> Yours very sincerely,
> Harry Pennell

Letters pertaining to Lieutenant. Pennell's marriage and his death in 1916 when he went down with his ship, HMS Queen Mary at the Battle of Jutland where he was serving as her navigating Commander in the First World War.

Francis Davies was witness to the extraordinary meeting of Roald Amundsen's ship Fram and Terra Nova

British and Norwegian expeditions meet in the Antarctic. The Terra Nova (with funnel) and the Fram, Captain Amundsen's ship, in the Bay of Whales.

Letters

Line Barn c/o Admiral Holland
Godalming Langley House, Chichester

July 28. 1916

Dear Mr Davies

It was very kind of you to write to me and I do thank you most sincerely for your kind letter.

When my husband was last home on leave he shewed me a tobacco jar which you had made while in Terra Nova — it had got a special place among his possessions & I have it now in my possession. That is only one of the ways in which he mentioned you, & I just told in to shew you it did not seem that you were quite a stranger & I try to feel content in knowing that my husband as a sailor, died the death he would have chosen — and I am intensely proud of him. But these days are not easy ones.

Dr Atkinson told me that you were going to write. With my thanks and best wishes for your safety,

Yours sincerely
Katie Pennell

Letters

Dec. 26th
Lone Barn
Godalming
Surrey

Dear Mr Davis

I am hoping to build a Church in C. Africa where I am working, in memory of my brother Commander Pennell who, as you know, went down in H.M.S. Queen Mary. I was talking to Dr Atkinson about it, + he said that he felt sure you would like to be asked to contribute towards the Church – I know that we are living in times when every 1ᵈ is of value. So I do not want you to feel that a big subscription is wanted. If you can spare 6ᵈ or 1/- + care to send it to Mrs Harry Pennell we shall be so pleased to have it, + value it as a gift from a friend of my brother. Mrs Pennell's address is on the top of this letter. Dr Atkinson only got a very few days leave from France. so I was more than lucky in seeing him. He looked very well – though a little tired from his journey. I have not met anyone else. though I had a line from Burton the other day. The Cross you made me is in Africa awaiting my return. which I hope will be early in Jan.

With best wishes for 1917
I remain
Yours sincerely
Winifred Pennell

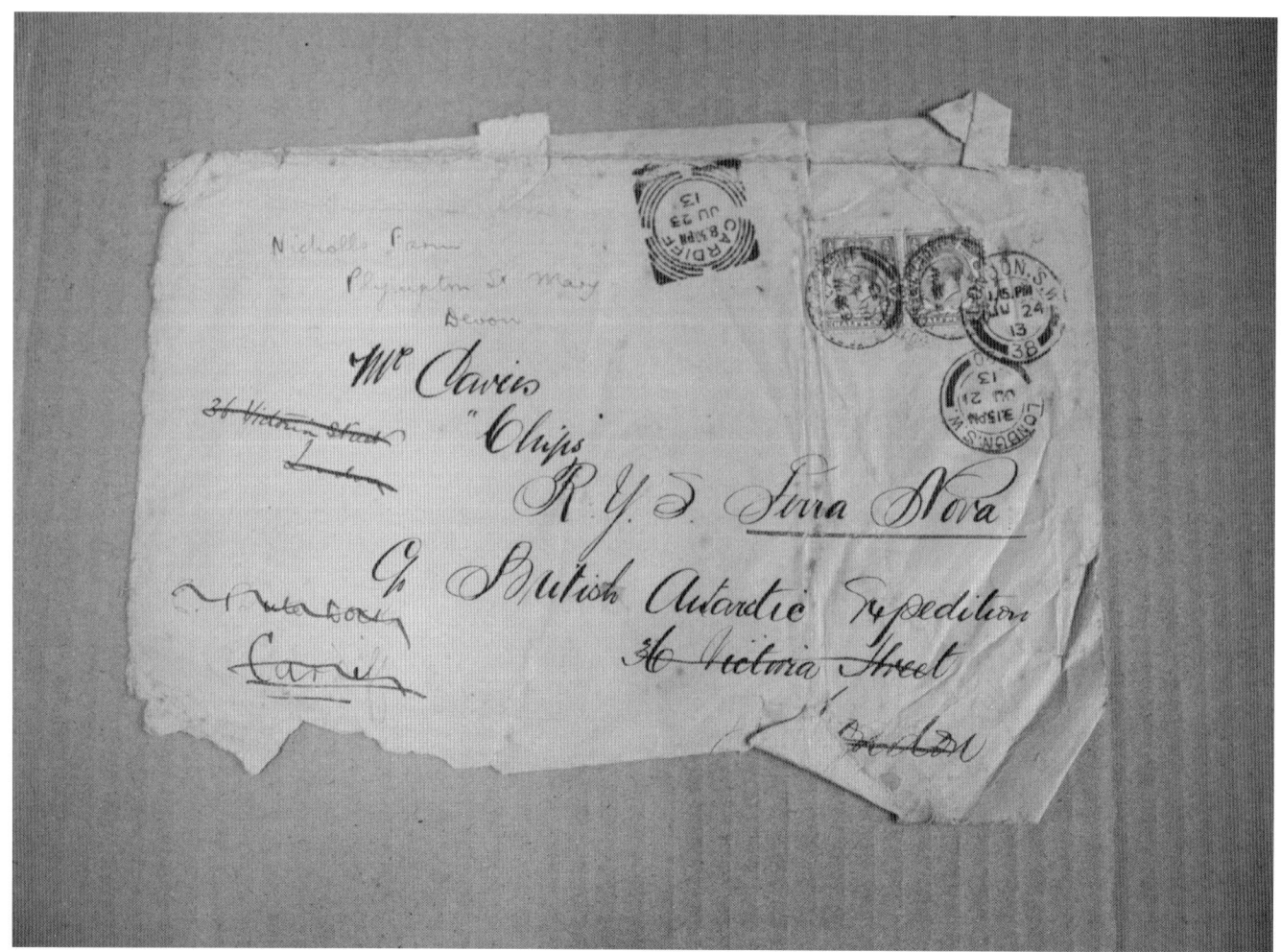

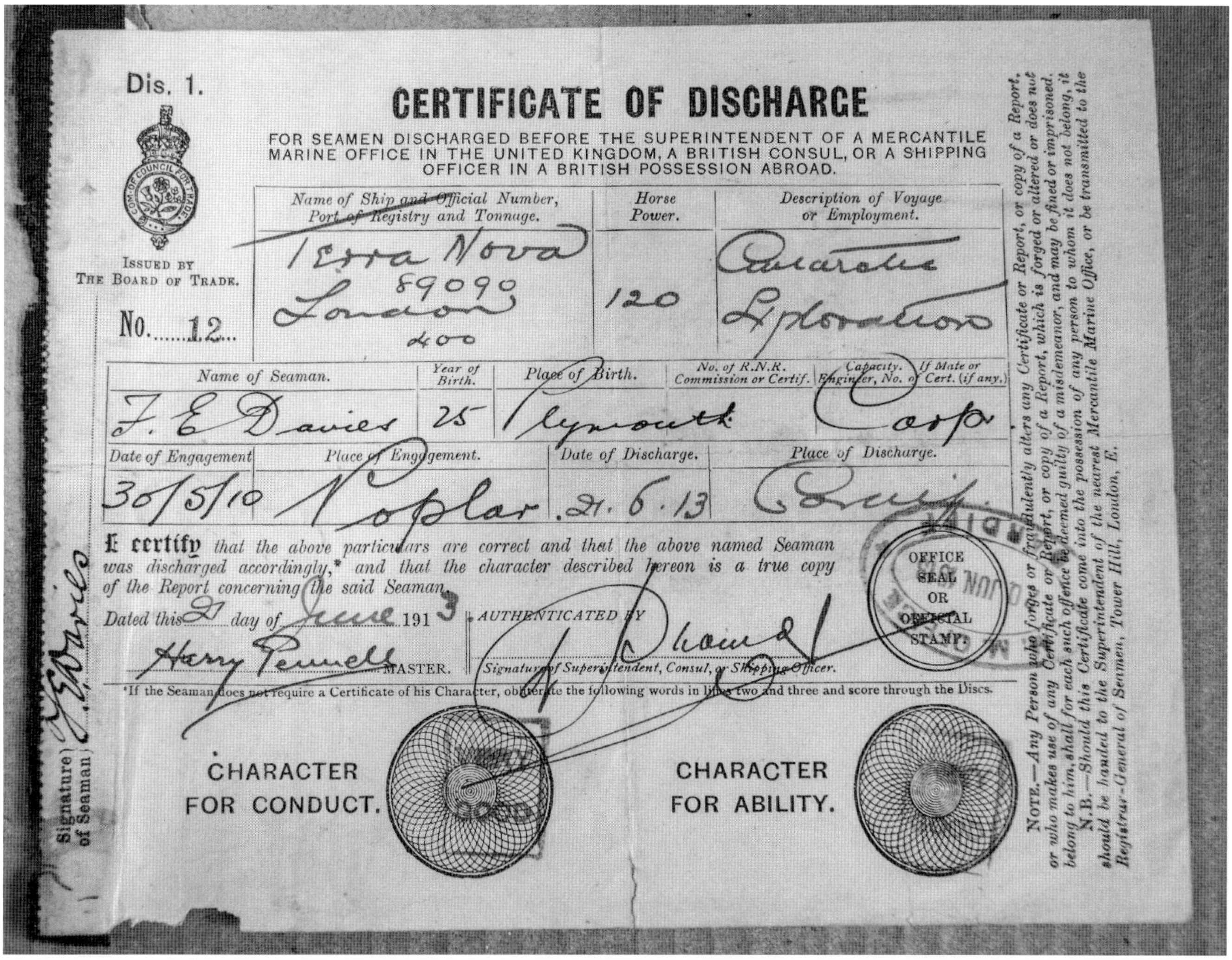

Dis. 1.

CERTIFICATE OF DISCHARGE

FOR SEAMEN DISCHARGED BEFORE THE SUPERINTENDENT OF A MERCANTILE MARINE OFFICE IN THE UNITED KINGDOM, A BRITISH CONSUL, OR A SHIPPING OFFICER IN A BRITISH POSSESSION ABROAD.

ISSUED BY THE BOARD OF TRADE.

No. 12

Name of Ship and Official Number, Port of Registry and Tonnage.	Horse Power.	Description of Voyage or Employment.
Terra Nova 89090 London 400	120	Antarctic Exploration

Name of Seaman.	Year of Birth.	Place of Birth.	No. of R.N.R. Commission or Certif.	Capacity. Engineer, No. of Cert.	If Mate or Cert. (if any.)
F. E. Davies	25	Plymouth		Carp	

Date of Engagement	Place of Engagement.	Date of Discharge.	Place of Discharge.
30/5/10	Poplar	21.6.13	Cardiff

I certify that the above particulars are correct and that the above named Seaman was discharged accordingly,* and that the character described hereon is a true copy of the Report concerning the said Seaman.

Dated this 21 day of June 1913

Harry Pennell MASTER.

AUTHENTICATED BY

Signature of Superintendent, Consul, or Shipping Officer.

OFFICE SEAL OR OFFICIAL STAMP.

*If the Seaman does not require a Certificate of his Character, obliterate the following words in lines two and three and score through the Discs.

CHARACTER FOR CONDUCT.

CHARACTER FOR ABILITY.

Signature of Seaman F. E. Davies

NOTE.—Any Person who forges or fraudulently alters any Certificate or Report, or copy of a Report, or who makes use of any Certificate or Report, or copy of a Report, which is forged or altered or does not belong to him, shall for each such offence be deemed guilty of a misdemeanor, and may be fined or imprisoned. N.B.—Should this Certificate come into the possession of any person to whom it does not belong, it should be handed to the Superintendent of the nearest Mercantile Marine Office, or be transmitted to the Registrar-General of Seamen, Tower Hill, London, E.

Davies' Certificate of Discharge from Terra Nova.

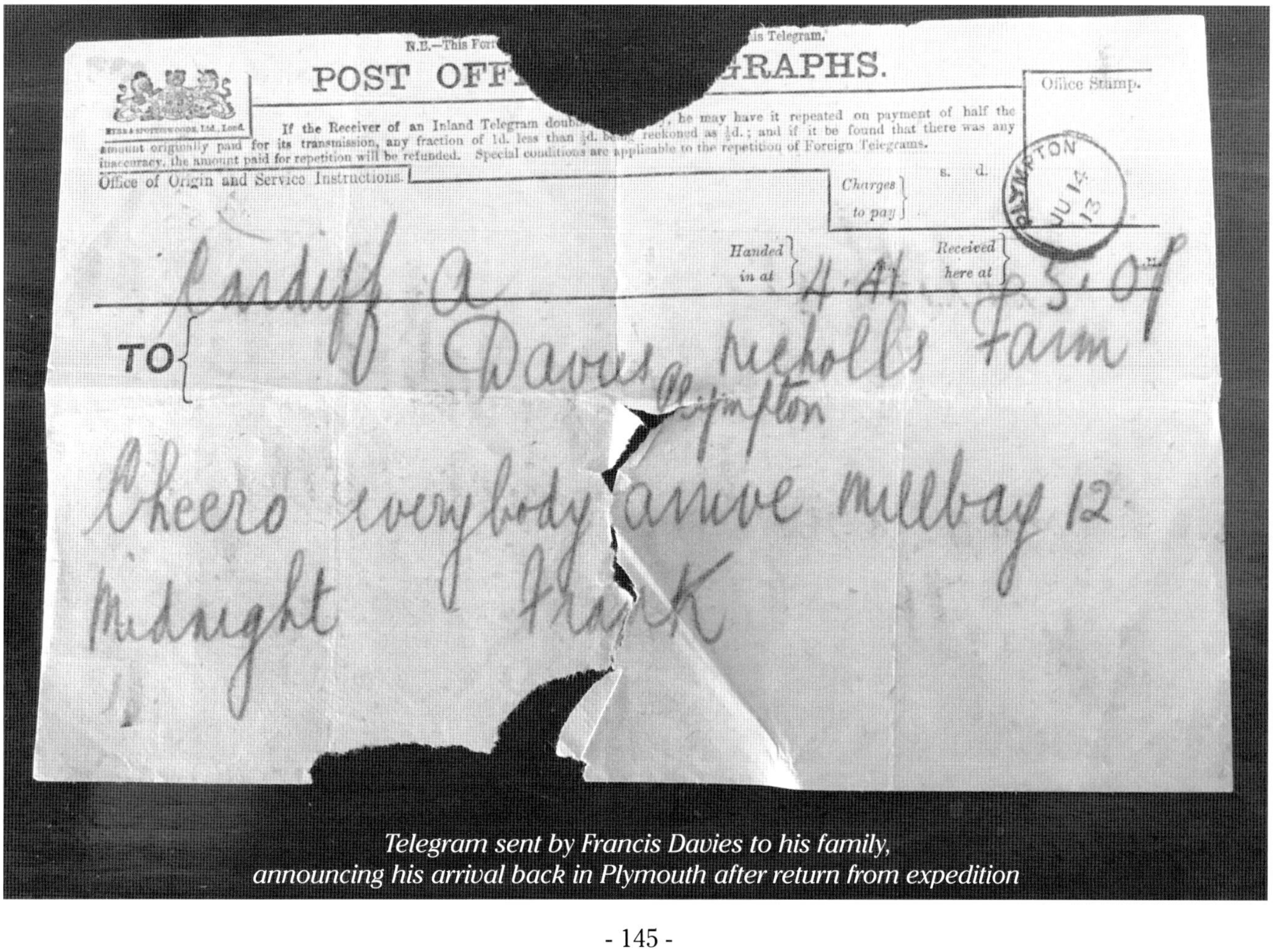

POST OFF... GRAPHS.

If the Receiver of an Inland Telegram doub... he may have it repeated on payment of half the amount originally paid for its transmission, any fraction of 1d. less than ½d. being reckoned as ½d.; and if it be found that there was any inaccuracy, the amount paid for repetition will be refunded. Special conditions are applicable to the repetition of Foreign Telegrams.

Office of Origin and Service Instructions.

Charges to pay s. d.

Office Stamp. PLYMPTON JU 14 13

Cardiff a Handed in at 4.41 Received here at 5.07

TO Davies Nicholls Farm Plympton

Cheero everybody arrive millbay 12 midnight Frank

Telegram sent by Francis Davies to his family, announcing his arrival back in Plymouth after return from expedition

Francis Davies' Career following the British Antarctic Expedition Terra Nova 1910 - 1913

Francis Davies returned to service in the Royal Navy.

During the First World War he served aboard HMS *Blanche* 1914 - 18 with the Grand Fleet and on mine laying.

In 1918 he served in HMS *Exmouth* and in 1919 went to Archangel in Russia where he was placed in charge of docking and repairs to all shipping operating in the White Sea and Severnaya Dvina River.

In 1919 - 20 he served in HMS *Sandhurst* during operations in the Baltic.

In 1920 he took early voluntary retirement from the Royal Navy.

August 1932 he passed the Board of Trade Examination for Certificate of Competency as Master (Steamship - Foreign going).

During the period 1927 - 34 he served in Royal Research Ships *Discovery II* and *William Scoresby* engaged in scientific work in the Southern Ocean regions. In *Discovery II* as 4th, 3rd and 2nd Officer and in *William Scoresby* as 2nd Officer and Navigator, Chief Officer and Navigator and as Master during Planktonic and Hydrographic Survey of the Humboldt Current off Chile and Peru in 1931.

He volunteered and served in the Second World War and was granted a Temporary Commission as a Lieutenant RNVR on 18 April 1940 and was appointed to HMS *Victory III* for special service, on 21 April 1940 proceeded to Harstad, Norway where he served with the naval detachment until evacuation on 8 June 1940. He was commended by the Rear Admiral, Narvik and the NOIC Harstad for his services in Norway.

On 17 July 1940 appointed to Boom Defence Department and transferred to RNR.

On 19 September 1940 appointed Assistant Boom Officer, Boston in charge of the Wash.

Later promoted to Lieutenant Commander RNVR his final appointment being Boom Defence Officer, Grimsby.

Lieutenant Commander F.E.C. DAVIES RNVR in later life

To Strive

To Seek

To Find

And Not To Yield

The National Memorial to Captain Scott and The Polar Party at Mount Wise, Devonport, Plymouth.

ANTARCTIC HERITAGE TRUST
— INSPIRING EXPLORERS —

Antarctic Heritage Trust is a New Zealand-based not-for-profit with a vision of 'Inspiring Explorers'. Through its mission to conserve, share and encourage the spirit of exploration the Trust cares for the expedition bases left behind by Antarctic explorers including, Carsten Borchgrevink, Captain Robert Scott, Sir Ernest Shackleton and Sir Edmund Hillary. In 2002, HRH Princess Anne launched the Trust's Ross Sea Heritage Restoration Project (RSHRP) in Antarctica.

This is the world's largest multi-year, multi-site heritage conservation programme. The Trust has engaged international heritage and conservation specialists from more than 14 countries to join the team in Antarctica to work in custom-built facilities in the most challenging heritage conservation environment on earth. Today the Trust has completed a major phase of conservation work on Ross Island, which included the conservation of our historic bases and many thousands of artefacts.

Sir Ernest Shackleton's 1908 base and its collection of more than 6,000 artefacts Captain Robert Falcon Scott's last expedition base at Cape Evans and its 11,500 artefacts Scott's first expedition base at Hut Point and 500 artefacts Hillary's (TAE/IGY) Hut at Scott Base and its 600 artefacts All four sites have a comprehensive monitoring and maintenance programme of work in place.

Antarctic Heritage Trust is now working on the conservation of Antarctica's first building. This historically significant site at Cape Adare is the only example left of humanity's first dwelling on any continent. Teams of professional conservators, with expertise in paper, timber, textile and metal conservation have either over-wintered at Scott Base or spent summers in the field working from the conservation laboratory. Collectively, they have conserved a staggering 20,000 individual artefacts including clothing, equipment and personal items.

The conservation work has led to a number of significant discoveries including 114-year old whisky found under Shackleton's hut, a notebook from surgeon and photographer George Murray Levick at Scott's Cape Evans hut as well as lost Ross Sea Party photographs. Conservators discovered a century old fruitcake (almost perfectly preserved) and a 118-year old watercolour amongst artefacts from Cape Adare. The Trust is proud to care for these early explorer bases and the remarkable legacy they left behind on behalf of the world. The Trust also shares the legacy of exploration through outreach programmes and encourages the spirit of exploration through expeditions to engage and inspire a new generation. More information about the Trust's conservation work and outreach programmes can be found at www.nzaht.org.

Scott's hut at Cape Evans

New Zealand Antarctic Heritage Trust
Outreach Programme Information

One of the Trust's strategic goals is to encourage young people to explore the world to educate and inspire them. A focus for the Trust is connecting young people with the spirit of exploration and the legacy the Trust cares for in Antarctica.

The Inspiring Explorers' Expeditions are one of the Trust's initiatives where young people from around the world have the opportunity to join the Trust's team on expeditions to explore and further understand the polar regions.

To date, the Trust has led four expeditions, including a journey to retrace Shackleton, Worsley and Crean's footsteps across South Georgia Island, honouring the epic crossing 100 years after the original journey.

Other expeditions have included an ascent of Mt Scott on the Antarctic Peninsula, a crossing of the Greenland ice cap to honour Fridtjof Nansen's first crossing 130 years earlier, and a trip for young New Zealanders to explore the Antarctic Peninsula.

As part of these experiences, each Inspiring Explorer shares their story to inspire others to go out and explore. Stories are shared through photography, short film, media, social media, blogging and public presentations. The Trust has produced a number of films about their expeditions.

The Trust is also involved in developing projects such as exhibitions and virtual reality to showcase the legacy it cares for in Antarctica with people all around the world.

The UK Antarctic Heritage Trust (UKAHT) is the UK's only organisation concerned with the preservation of British heritage in Antarctica. The UK has a long and distinguished history of endeavour in Antarctica, from its first discovery two centuries ago to cutting edge Polar science today. The legacy of the early pioneers lives on, through the stories that are told, through the artefacts preserved in museums but also through the modest huts which still remain in Antarctica.

These historic huts erected by the protagonists of some of these heroic tales are the real and physical evidence of some of the world's most enduring stories of endurance, heroism, triumph and tragedy. Preserving these huts is an endeavour which can, at times, be just as demanding but doing so ensures that legacy lives on into the future to inspire future generations.

Designated and protected as monuments under the Antarctic Treaty and Environmental Protocol Scott's huts, and those which came afterwards on the Antarctic Peninsula, are our Antarctic memory. Whilst few people will ever see them in person, their preservation means that the endeavours and sometime sacrifices of those early explorers and scientists who prepared the way for the rest of us, will never be forgotten.

The labours of Francis Davies in constructing the hut at Cape Evans echo throughout the century that followed. Doing anything in Antarctica is demanding, but designing and building shelter to sustain people through the worst of conditions is a huge responsibility as well as technically challenging. The problems he faced over a century ago are the same as those faced by generations since. Even the most modern state-of-the-art Antarctic buildings need to withstand relentless winds, snow and ice and solve the same problems.

In caring for historic huts, we have a unique perspective on the decisions made by those who designed and built them; the innovation, the problem solving, the compromises and the creativity. Sometimes they used the most advanced materials of the day, contrasting these with techniques that wouldn't have been out of place in a medieval timber-framed building. Peeling back the layers and understanding these buildings offers insight into the men who built them and inhabited them - their skills, creativity and sheer will to conquer the worst nature has to offer.

Caring for these huts, whether they are Scott's in the Ross Sea, or the early science bases on the Antarctic Peninsula, is a huge privilege and responsibility and can only happen if organisations like the UKAHT and sister trust in New Zealand are supported in order to continue to carry out their vital work.